水利工程专业英语

肖　毅　主编

人民交通出版社股份有限公司

北　京

内 容 提 要

本教材从水利工程基础专业术语英文释义入手，通过各类水利工程的设计、施工及运行等方面英文专业论述，系统梳理水利工程专业术语的英文释义及科技论文写法。书中内容主要分为6个单元，共46篇课文与相关阅读材料，涉及水利工程领域各个专业，包括水文与水资源、水利水电工程、港口航道工程以及水利施工管理等相关内容。

本书可作为高等院校水利类专业本科生和研究生的专业英语教材或课外阅读材料，也可供从事相关专业的科技人员、工程技术人员、管理人员和教师查阅使用。

图书在版编目（CIP）数据

水利工程专业英语 / 肖毅主编. — 北京：人民交通出版社股份有限公司，2019.8

ISBN 978-7-114-15800-1

Ⅰ. ①水… Ⅱ. ①肖… Ⅲ. ①水利工程—英语—高等学校—教材 Ⅳ. ① TV

中国版本图书馆 CIP 数据核字（2019）第 189151 号

Shuili Gongcheng Zhuanye Yingyu

书 名：水利工程专业英语
著 作 者：肖 毅
责任编辑：闫吉维
责任校对：孙国靖 龙 雪
责任印制：刘高彤
出版发行：人民交通出版社股份有限公司
地 址：（100011）北京市朝阳区安定门外外馆斜街3号
网 址：http://www.ccpress.com.cn
销售电话：（010）59757973
总 经 销：人民交通出版社股份有限公司发行部
经 销：各地新华书店
印 刷：北京鑫正大印刷有限公司
开 本：787×1092 1/16
印 张：13.25
字 数：303千
版 次：2019年8月 第1版
印 次：2019年8月 第1次印刷
书 号：ISBN 978-7-114-15800-1
定 价：39.00元

前　言

本书围绕水利工程学科进行选材，从水利工程基础课程专业术语入手，由简至繁，通过对各类水利工程的设计、施工及运行等方面的英文论述，系统梳理水利工程专业术语英文释义，与国外本学科专业课程英文教程接轨；同时将水利工程世界前沿发展趋势融入书中，为培养具有国际视野的工科复合型人才提供语言基础。图文并茂、通俗易懂是本书的最大特色，全书分为6个单元，由23篇课文、23篇阅读材料及科技英语阅读写作方法组成，内容涵盖水资源及水文循环、土力学、水利水电工程、港口工程、航道工程、典型水利工程案例、工程施工及工程经济、科技英语阅读与写作等方面。

本书由重庆交通大学河海学院教师合作编写。河海学院肖毅副教授任主编，李文杰教授、刘洁副教授任副主编，全文由肖毅统稿，河海学院杨胜发、胡江教授主审。教材编写大纲由编写人员集体讨论确定。本书在编写过程中得到了重庆交通大学以及各位编者的大力支持。由于时间仓促，加之编者水平有限，对于书中不妥之处，恳请读者批评指正，提出改进意见。

编　者

2019年6月

前　言

目　录

UNIT I INTRODUCTION TO HYDRAULIC ENGINEERING

Lesson 1 Water Resources

Water is an indispensable resource. Natural water resources are sources of water that are used for domestic, agricultural, industrial, recreational, and environmental activities. All living things require water to survive and thus it is told "water means live and live means water". The Earth's surface consists of 25% of land and 75% of water, and all creatures need water to survive. Among the surface water, 97% is saline while only 3% is freshwater. What's more, over two-thirds of the freshwater exist in the form of glaciers or ice caps. However, most of the rest of the freshwater which is unfrozen proves to be groundwater, and the final remainder is above ground or in the air. Fresh water is a renewable resource, yet the world's supply of groundwater is steadily decreasing, with depletion occurring most prominently in whole world which is threatening our survival near future (Fig. 1.1).

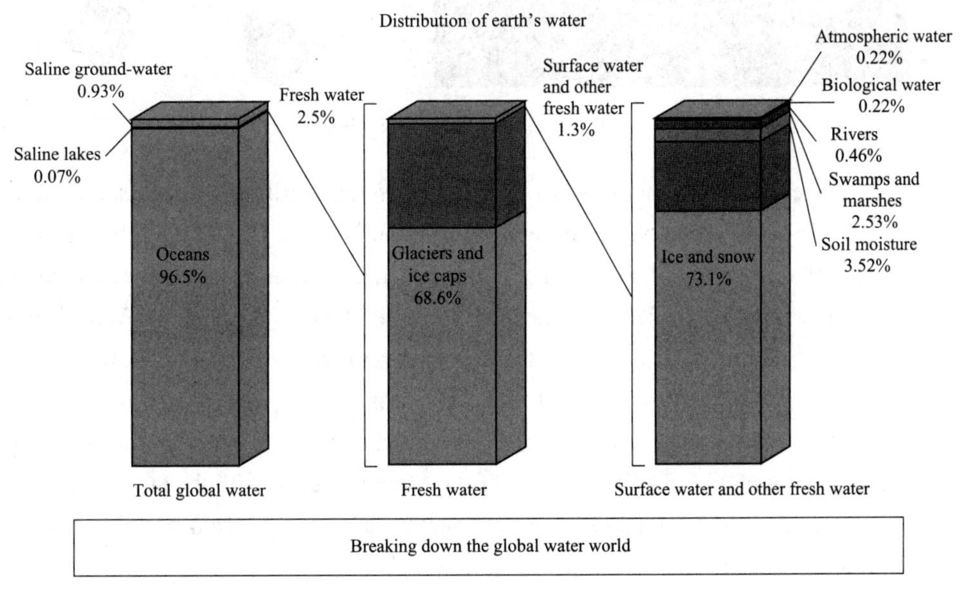

Figure 1.1 A graphical distribution of the locations of water on the Earth

1. Chemical and physical properties of water

H_2O, a polar inorganic compound, represents as liquid with attributes of tasteless, odorless, and nearly colorless with a little bit blue under normal condition. It is the simplest hydrogen chalcogenide which is described as a universal solvent because of its ability to dissolve many substances, e.g., many salts, sugar, simple alcohols, proteins, polysaccharides, DNA, oxygen, and carbon dioxide.

Water molecule is a polar molecule with an electrical dipole because the oxygen atom has a higher electronegativity than hydrogen atoms, which means the oxygen atom carries a slight negative charge, while hydrogen atoms are slightly positive. Besides, each water molecule contains one oxygen atom and two hydrogen atoms connected by covalent bonds (Fig.1.2). More importantly, it is the only common substance that can exist as solid, liquid, and gas in normal terrestrial conditions. Water can be referred to the liquid state of a substance that exists at standard ambient temperatures and pressures, its solid state (ice), or its gaseous state (steam or water vapor). It also exists in the forms of ice packs and icebergs, fog, glaciers, aquifers, atmospheric humidity, fog, and dew, which are the main components of the Earth's streams situated in difference locations.

Figure 1.2 Model of hydrogen bonds

Under different circumstances, water will have different attributes. When standard pressure is 1 atm, water is in the form of liquid between 0 ℃ (32 ℉) and 100 ℃ (212 ℉). However, the melting point will change to -5℃ at 600 atm and -22℃ at 2100 atm, which is used as the principle to explain the movement of glaciers, ice skating, and buried lakes of Antarctica. What's more, the melting point will rapidly increase again, leading to several exotic forms of ice that do not exist at lower pressures when pressure is higher than 2100 atm.

The boiling point will also change when pressure changes. For instance, the boiling point is 374℃ at 220 atm but 68℃ (154°F) at the top of the Mount Everest with the atmospheric pressure of 0.34 atm. The effect of this change is applied on pressure cooking, steam engine design, deep-sea hydrothermal vents, etc.

Water attributes change under extreme conditions. When the pressure is extremely low (lower than 0.006 atm), water cannot exist in liquid form and passes directly from solid to gas by sublimation, which is used for freeze drying of food. When the pressure is extremely high (higher than 221 atm), the liquid and gas states are no more distinguishable, a new state called supercritical steam replaces.

Another characteristic of water is that it becomes less dense as it freezes compared with most liquids. The maximum density of water in its liquid form (at 1 atm) is 1000 kg/m^3 (62.43 lb/cu·ft), which occurs at 3.98 ℃ (39.16℉) while the density of ice is 917 kg/m^3 (57.25 lb/cu·ft). Thus, through calculation, water expands 9% in volume as it freezes, which reflects the fact that ice floats on liquid water.

2. Surface water

Surface water refers to water locates in river, lake, or freshwater wetland. It is naturally refilled and lost by precipitation and discharge to the oceans, evapotranspiration, groundwater recharge, and evaporation. Although precipitation is the only natural input to any surface water system, many other factors also influence the total quantity of water available at any given time and proportions of water loss in the system, for example, wetlands and artificial reservoirs, the runoff characteristics of the land in the watershed, storage capacity in lakes, the timing of the precipitation, the permeability of the soil beneath these storage bodies, and local evaporation rates.

However, human activities can have large or even destructive impacts on these factors. By constructing reservoirs or draining wetlands, human can control the storage capacity of the surface water systems. Besides, by paving areas and channelizing the streamflow, human can increase runoff quantities and velocities.

The total quantity of water available at any given time is an important issue to considerate. Some water users have a desultory need for water, farms, for example, only need water when planting crops, which maybe only one season in the whole year. To satisfy the farm's need for water, a surface water system may require a large storage capacity to gather water during the whole year to release it in a short period of time. For other users who need continuous supply of water such as power plants, the storage capacity of the surface water system only needs to satisfy the minimum requirement of the average streamflow.

3. Under river flow

During the flow process of a river, the total volume of water transporting downstream normally consists of the visible free water flow and the flow across rocks and sediments that underlie the river, whose floodplain is called the hyporheic zone. For rivers in large valleys, the latter flow may largely exceed the visible flow. The hyporheic zone often plays the roles of forming a dynamic interface between surface water and groundwater from aquifers and exchanging flow that may be fully charged or depleted between rivers and aquifers, which are particularly significant in karst areas where pot-holes and underground rivers are common.

4. Groundwater

Groundwater can be referred to the freshwater located in the subsurface pore space of soil and rocks and the water flows within aquifers below the water table. The critical difference between groundwater and surface water is that, compared to input, the volume of groundwater storage is much larger than surface water because of the former's slow rate of turnover. This difference let human to use groundwater unsustainably for a long time if there is no serious issue. However, the average rate of seepage above a groundwater source is the upper bound for average consumption of water from that source in the long term. The natural input and output of groundwater is the seepage from surface water and springs or seepage to the oceans, respectively. Therefore, if the surface water source is subject to substantial evaporation, the groundwater source below may become saline. This circumstance may happen artificially under irrigated farmland or naturally under endorheic bodies of water. Soil salinization may also happen in coastal areas because of the reverse of direction of seepage to ocean due to human use. What's more, the pollution that results from human activities may cause the loss of groundwater which can be stopped by building reservoirs or detention ponds.

5. Environmental impact

In 2025, water shortage problems will be more obvious among poorer areas with limited resources and rapid population growth, such as the Middle East, parts of Asia, and Africa. Besides, at that time, to provide clean water and adequate sanitation, large urban and periurban areas will need to construct new infrastructures, which may lead to growing conflicts between government and agricultural water users, who currently consume most of the water. Compared with these poorer areas, more developed areas of North America, Europe, and Russia will not suffer the threat, not only because of their relative wealth but also because that their population better can align with available water resources.

6. Water pollution

One of the major concerns today is water pollution and many countries have tried hard to find solutions to solve this problem. Among all pollutants threatening water supplies, the most widespread one is the discharge of raw sewage, sludge, garbage, or even toxic pollutants into natural water, which is most common in underdeveloped countries and some quasi-developed countries such as India, Iran, and Nepal. Treated sewage can form sludge, which may be placed in landfills, spread out on land, incinerated or dumped at sea. Except for sewage, nonpoint source pollution such as agricultural runoff, urban storm water runoff, and chemical wastes are also a major source of pollution in some countries.

7. Climate change

The misuse of water resources worldwide significantly impacts the climate change because of the close connections between the climate and hydrological cycle. Increased hydrologic variability and change in climate has and will continue to have a profound impact on the water sector through the hydro- logic cycle, water availability, water demand, and water allocation at the global, regional, basin, and local levels. Rising temperatures will increase evaporation and further lead to increase in precipitation. Besides, it will also affect water quality by increasing eutrophication and other not well-understood reasons presently. Climate change could also result in much demand for garden sprinklers, swimming pools, farm irrigation, etc.

New Words and Phrase

1.indispensable *adj.* 必要的
2.saline *adj.* 含盐的，咸的
3.freshwater *n.* 淡水
4.groundwater *n.* 地下水
5.depletion *n.* 消耗，耗减
6.hydrogen *n.* 氢
7.chalcogenide *n.* 氧属（元素）化物
8.dissolve *vi.* 溶解
9.substance *n.* 物质，材料
10.alcohol *n.* 酒精
11.protein *n.* 蛋白质
12.dioxide *n.* 二氧化物
13.polar molecule 极性分子
14.charge *n.* 电荷
15.atom *n.* 原子
16.covalent bond 共价键
17.iceberg *n.* 冰山
18.aquifer *n.* 地下蓄水层
19.atmospheric *adj.* 大气的
20.dew *n.* 露水
21.hydrothermal *adj.* 水热的
22.sublimation *n.* 升华作用
23.wetland *n.* 湿地
24.reservoir *n.* 水库
25.runoff *n.* 径流
26.power plants 发电厂房
27.endorheic *adj.* 背阴的
28.detention pond 蓄水池
29.sanitation *n.* 卫生设备
30.sewage *n.* 污水，下水道
31.sludge *n.* 污泥
32.incinerate *vi.* 焚化
33.nonpoint source 非点源
34.hydrological cycle 水循环
35.allocation *n.* 分配额度

Notes

1.Natural water resources are sources of water that are used for domestic, agricultural, industrial, recreational, and environmental activities.

自然水资源是人类生活、农业、工业、娱乐以及环境活动用水的主要来源。

source of 指来源，其后一般接名词或者名词性短语：

It is the source of castor oil, which has a wide variety of uses.

be used for 一般用于被动语态，表示“（某物）被用于……”，其后接名词、代词或动名词：

The 2D numerical model is used for the simulation of the flow pattern in the upstream of the Yangtze River.

2.It also exists in the forms of ice packs and icebergs, fog, glaciers, aquifers, atmospheric humidity, fog, and dew, which are the main components of the Earth's streams situated in difference locations.

它主要以浮冰、冰山、冰河、地下蓄水层、大气湿度层、雾以及露水等形式存在，而它们是构成地球各类河流的主要部分。

in the form of　以……的形式

The amount of water vapor passing over the United States every day is approximately 152000 million m^3, and of this approximately 10% falls as precipitation in the form of rain, snow, hail, or sleet.

component of　组成，成分

What is the main component of hydrological cycle?

3.Compared with these poorer areas, more developed areas of North America, Europe, and Russia will not suffer the threat, not only because of their relative wealth but also because that their population better can align with available water resources.

相对于贫困国家，北美、欧洲以及俄罗斯等发达国家将不会受到水源短期问题的威胁，不仅是因为他们相对富有，更重要是由于他们的人口能与可利用水资源匹配一致。

not only…, but also…　不仅…… 而且……，其中 also 可以省略；放在句首可倒装

Not only do the professors have their own ideas on this issue, but the students have theirs too.

Comprehensive Exercises

I.Answer the following questions on the text.

1.What is the earth's surface water consist of?

2.What is the characteristics of water?

3.How many types for the water attributes, what are they?

4.What is the impact of the misuse of water resources on climate change?

II.Fill the most appropriate words or phrases in the correct forms in the blanks from the list below.

survive	consist of	universal	refer to	in the form of
be used for	be subject to	align	hydrological cycle	allocation

1.The imposing bridges that ________ancient times are arched structures of heavy masonry, usually stone or brick.

2.Gravity dams ________solid concrete or masonry dams of roughly triangular cross section, which depend primarily on their own weight and cohesion with the foundation for stability.

3.There is no________solution for developing a catchment or farm water management plan that suits all the varied landscapes and soil types in the wheat belt.

4.Energy________both pressure and kinetic is utilized by the wheel.

5.Linings________plain or reinforced concrete, cement mortar, asphalt, brick, stone, buried synthetic membranes, or compacted earth materials.

6.Land protected by levees______excess water as a result of seepage from the river through or under the levee.

7.It's supposed to______the interests of the individual employee with the bank and the shareholders.

8.Changes of______ are controlled by natural and human activities.

9.Communications satellites______international living transmission throughout the world.

10.The_______material costs is based on the quantity of each type of material used for each specific item.

III.Translate the following sentences into Chinese from the text.

1.To satisfy the farm's need for water, a surface water system may require a large storage capacity to gather water during the whole year to release it in a short period of time.

2.During the flow process of a river, the total volume of water transporting downstream normally consists of the visible free water flow and the flow across rocks and sediments that underlie the river, whose floodplain is called the hyporheic zone.

3.The misuse of water resources worldwide significantly impacts the climate change because of the close connections between the climate and hydrological cycle. Increased hydrologic variability and change in climate has and will continue to have a profound impact on the water sector through the hydro- logic cycle, water availability, water demand, and water allocation at the global, regional, basin, and local levels.

Reading Material Water Uses

1.Domestic

According to estimation, 8% of water use worldwide is for domestic purposes, which include cooking, laundry, toilet flushing, drinking water, bathing, gardening, and cleaning. Excluding water for gardening, basic domestic water requirement is at around 50 L per person per day according to estimation. Drinking water, more special than other uses, refers to water that is of sufficiently high quality so that it can be consumed or used without risk of harmless, which can also be called potable water. In most developed countries, the water provided for domestic, commercial, or industrial purposes is at drinking water standard, although only a small proportion is consumed for food preparation.

2.Agricultural

Compared with domestic use, agricultural use of water occupies a large amount of water use. According to an estimation, 70% water use worldwide is irrigation, with 15%-35% of irrigation withdrawals being unsustainable. To satisfy one person's daily food need, 2000-3000 L of water will be consumed, which is a considerable amount compared to drinking 2-5L. Therefore, according to calculation, if enough food is provided for over 7 billion people now on the earth, the water required could fill a canal 10 m deep, 100 m wide, and 2100 km long.

3. Industrial

Another major water use is industrial, which occupies 22% worldwide water use. Major application of industrial water use includes thermoelectric power plants, manufacturing plants, ore and oil refineries, and hydroelectric dams. Water in these processes can play the roles of cooling, solvent, and chemical reagents. Although water withdrawal may be high in some industries, these consumption are still much lower than that of agriculture. Industrial water uses can also include renewable power generation. Hydro-electric power derives energy from the force of water flowing downhill, driving a turbine connected to a generator, which is a low cost, nonpolluting, renewable energy source. It can be used for load following and can provide continuous power.

Another industrial water use is pressurized water, which is often used in water blasting, water jet cutters, and high-pressure water guns. These uses not only work well and safe but also make no harm to the environment. Cooling of machinery from overheating is also an industrial water use, which only occupies a small proportion of water consumption. In many large-scale industrial processes such as oil refining, fertilizer production, and other chemical plants, thermoelectric power production, and natural gas extraction from shale rock, water is also being used. Discharge of untreated water coming from these processes is pollution, which includes discharged solutes (chemical pollution) and increased water temperature (thermal pollution). These industrial water use mentioned above normally require pure water and apply a variety of purification techniques both in water supply and discharge. Most of the pure water is produced on-site from natural freshwater or municipal gray water.

4. Recreation

Sustainable management of water resources including protection of aquatic ecosystems, flood protection, adequate sanitation, provision of safe and reliable supplies for drinking water, irrigation, etc. Poses huge challenges world-wide. Recreational water use is generally a very small but increasing percentage in total water use. Normally, recreational water use is tied to reservoirs. If a reservoir is kept fuller than it would otherwise be for recreation, then the water retained could be categorized as recreational usage. Common recreational usage includes whitewater boating, angling, water skiing, and swimming. Recreational water use is usually non-consumptive. However, it is not clear whether recreational irrigation has an obvious influence on water resources till now because of the inapplicability of reliable data. Besides, many golf courses use either primarily or exclusively treated effluent water, which will not affect potable water. For example, water retained in a reservoir to allow boating in the late summer is not available to farmers during the spring planting season. Water released for whitewater rafting may not be available for hydroelectric generation during the time of peak electrical demand.

5. Environmental

Environmental water use is also a small but increasing percentage of total water use. The source of environmental water includes water stored in retention tank and released for environmental purposes or water retained in waterways through regulatory limits of abstraction. Then the usages of environmental water include fish ladders, water releasing from reservoirs timed to help fish spawn, restore more natural flow regimes, watering of natural or artificial wetlands, and create wildlife habitat. Similar to recreational usage, environmental usage is non-consumptive but may influence other users. For example, water released from a reservoir to help fish spawn may not be available to farms upstream, and water retained in a river to maintain waterway health would not be available to water abstractors downstream.

Lesson 2 Physical Properties and Equations of Hydraulics

As a natural science, the variability of river processes must be examined through the measurement of physical parameters. This passage first describes dimensions and units, physical properties of water, kinematics of flow, the equation of continuity, and the equation of motion.

1.Dimensions and units

Physical properties are usually expressed in terms of the following fundamental dimensions: mass (M), length (L), time (T), and temperature (T°). Throughout the text, the unit of mass is preferred to the corresponding unit of force. The fundamental dimensions are measurable parameters that can be quantified in fundamental units.

In the SI system of units, the fundamental units of mass, length, time, and temperature are the kilogram (kg), the meter (m), the second (s), and degrees Kelvin (°K). Alternatively, the Celsius scale (°C) is commonly preferred because it refers to the freezing point of water at 0°C and the boiling point at 100°C.A Newton (N) is defined as the force required for accelerating 1 kg at 1 m/s^2. Knowing that the acceleration that is due to gravity at the Earth's surface, g, is 9.81 m/s^2, we obtain the weight of a kilogram from Newton's second law: $F = mass \times g = 1\ kg \times 9.81\ m/s^2 = 9.81\ N$. The unit of work (or energy) is the joule (J), which equals the product of 1 N × 1 m. The unit of power is a watt (W), which is 1 J/s. In the English system of units, the time unit is a second, the fundamental units of length and mass are, respectively, the foot (ft), equal to 30.48 cm, and the slug, equal to 14.59 kg. The force required for accelerating a mass of one slug at 1 ft/s^2 is a pound force (lb). Throughout this text, a pound refers to a force, not a mass. The temperature, in degrees Celsius, TC°, is converted to the temperature in degrees Fahrenheit, TF°, by TF° = 32.2 °F + 1.8 TC°.

2.Properties of water

Mass density of water, ρ. The mass of water per unit volume is referred to as the mass density ρ. The maximum mass density of water at 4 °C is 1000 kg/m^3 and varies slightly with temperature. In comparison, the mass density of sea water is 1025 kg/m^3 and, at sea level, the mass density of air is 1.29 kg/m^3 at 0 °C. The conversion factor is 1 slug/ft^3.

Specific weight of water γ. The gravitational force per unit volume of fluid, or simply the fluid weight per unit volume, defines the specific weight γ. At 10°C, water has a specific weight, γ = 9810 N/m^3 or 62.4 lb/ft^3 (1 lb/ft^3 =157.09 N/m). Specific weight varies slightly with temperature. Mathematically, the specific weight γ equals the product of the mass density ρ times the gravitational acceleration g = 32.2 ft/s^2 = 9.81 m/s^2:

$$\gamma = \rho g \tag{1.1}$$

Dynamic viscosity μ. As a fluid is brought into deformation, the velocity of the fluid at any boundary equals the velocity of the boundary. The ensuing rate of fluid deformation causes a shear stress τ_{zx} that is proportional to the dynamic viscosity μ and the rate of deformation of the fluid, dv_x /dz:

$$\tau_{zx} = \mu \frac{dv_x}{dz} \tag{1.2}$$

Kinematic viscosity ν. When the dynamic viscosity of a fluid μ is divided by the mass density ρ of the same fluid, the mass terms cancel out. This results in kinematic viscosity ν with dimensions L^2/T, decreasing with temperature. The viscosity of clear water at 20 °C is 1 centistokes = 1 cs = 0.01 cm^2/s = 1 × 10^{-6} m^2/s. The conversion factor is 1 ft^2/s = 0.0929 m^2/s.It is important to remember that both the density and the viscosity of water decrease with temperature. The maximum water density is found at 4 °C, and water either colder or warmer than 4 °C will be found near the surface. The density of ice increases as the temperature decreases. This causes the ice cover to crack during cold nights and expand to apply large forces on the banks of lakes, reservoirs, and wide rivers during warm days.

3.River flow kinematics

Flow kinematics describes fluid motion in terms of velocity and acceleration. In rivers, two orthogonal coordinate systems are common: (1) global right-hand Cartesian (x, y, z) systems, with x in the main downstream direction, y in the lateral direction to the left bank, and z upward; and (2) local cylindrical (r, θ, z) systems, in which r is the river radius of curvature in a horizontal plane.

The rate of change in the position of a fluid element is a measure of its velocity. Velocity is defined as the ratio between the displacement ds and the corresponding increment of time dt. Velocity is a vector quantity v that varies in both space (x, y, z) and time t. Its magnitude v at a given time equals the square root of the sum of squares of its orthogonal components:

$$v = \sqrt{v_x^2 + v_y^2 + v_z^2} \tag{1.3}$$

where $v_x = \mathrm{d}x/\mathrm{d}t$, $v_y = \mathrm{d}y/\mathrm{d}t$, and $v_z = \mathrm{d}z/\mathrm{d}t$.

A line tangent to the velocity vector at every point at a given instant is known as a streamline. The path line of a fluid element is the locus of the element through time, e.g., the path followed by a single buoy on a river. A streak line is defined as the line connecting all fluid elements that have passed successively at a given point in space, e.g., instantaneous position of all buoys released over time from a single point on a river.

4.Conservation of mass

The equation of continuity, or law of conservation of mass, states that mass cannot be created nor destroyed. The continuity equation can be written in either differential form, discussed in this section, or integral form.

In differential form, consider the infinitesimal control volume in Fig.1.3 filled with a fluid .The difference between the mass fluxes entering and leaving the differential control volume equal the rate of increase of internal mass. For instance, in the x direction, the net mass flux leaving the control volume is $[(\partial\rho_{\mathrm{m}}\, v_x)/(\partial x)]$ dx times the area dydz. The change in internal mass is $(\partial\rho_{\mathrm{m}}/\partial t)$ dxdydz. Repeating the procedure in the y and the z directions yields the following differential relationships:

Cartesian coordinates (x, y, z):

$$\frac{\partial\rho_{\mathrm{m}}}{\partial t} + \frac{\partial}{\partial x}(\rho_{\mathrm{m}} v_x) + \frac{\partial}{\partial y}(\rho_{\mathrm{m}} v_y) + \frac{\partial}{\partial z}(\rho_{\mathrm{m}} v_z) = 0 \tag{1.4}$$

Cylindrical coordinates (r, θ, z):

$$\frac{\partial\rho_{\mathrm{m}}}{\partial t} + \frac{1}{r}\frac{\partial}{\partial r}(\rho_{\mathrm{m}} r v_r) + \frac{1}{r}\frac{\partial}{\partial\theta}(\rho_{\mathrm{m}} v_\theta) + \frac{\partial}{\partial z}(\rho_{\mathrm{m}} v_z) = 0 \tag{1.5}$$

For the particular case in which sediment diffusion is not significant, the conservation of solid mass is also defined after ρ_{m} is replaced with C_{v}. For sediment transport problems in which turbulent diffusion and dispersion are significant, sediment continuity equation, including turbulent-mixing coefficients, should be used. For homogeneous incompressible suspensions without settling, the mass density is independent of space and time (ρ_{s}, ρ, ρ_{m} = const); consequently $\partial\rho_{\mathrm{m}}/\partial t = 0$ and the divergence of the velocity vector in Cartesian coordinates must be zero, i.e.

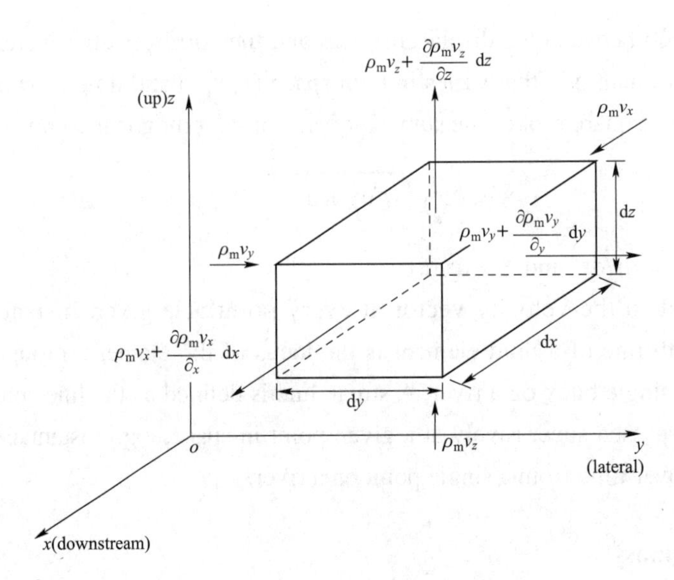

Figure 1.3 Infinitesimal element of a fluid

$$\frac{\partial v_x}{\partial x}+\frac{\partial v_y}{\partial y}+\frac{\partial v_z}{\partial z}=0 \tag{1.6}$$

When dealing with open-channel flows at low sediment concentrations, we can neglect compressibility effects, and we find that Eq. (1.6) is applicable.

5. Equations of motion

The analysis of fluid motion results from the application of forces on a fixed control volume. Given that the force F equals the product of mass m and acceleration a, the approach for fluids of mass density $\rho = (m/\forall)$ stems from $a = (F/m) = (F/\rho\forall)$. The forces acting on a Cartesian element of fluid and sediment (dx, dy, dz) are classified as either internal forces or external forces. The internal accelerations, or body forces per unit mass, acting at the center of mass of the element are denoted by g_x, g_y, and g_z. The external forces per unit area applied on each face of the element are subdivided into normal- and tangential-stress components. The normal stresses σ_x, σ_y, and σ_z are designated as positive for tension. Six shear stresses, τ_{xy}, τ_{yx}, τ_{xz}, τ_{zx}, τ_{yz}, and τ_{zy}, with two orthogonal components are applied on each face, as shown in Fig. 2.2. The first subscript indicates the direction normal to the face, and the second subscript designates the direction in which the stress is applied. The identities $\tau_{xy} = \tau_y$, $\tau_{xz} = \tau_{zx}$, and $\tau_{yz} = \tau_{zy}$ result from the sum of moments of shear stresses around the centroid.

The cubic element in Fig. 1.4 is considered in equilibrium when the sum of the forces per unit mass in each direction, x, y, and z, equals the corresponding Cartesian acceleration components a_x, a_y, and a_z:

$$a_x = g_x + \frac{1}{\rho_m}\frac{\partial \sigma_x}{\partial x} + \frac{1}{\rho_m}\frac{\partial \tau_{yx}}{\partial y} + \frac{1}{\rho_m}\frac{\partial \tau_{zx}}{\partial z}$$

$$a_y = g_y + \frac{1}{\rho_m}\frac{\partial \sigma_y}{\partial y} + \frac{1}{\rho_m}\frac{\partial \tau_{xy}}{\partial x} + \frac{1}{\rho_m}\frac{\partial \tau_{zy}}{\partial z} \tag{1.7}$$

$$a_z = g_z + \frac{1}{\rho_m}\frac{\partial \sigma_z}{\partial z} + \frac{1}{\rho_m}\frac{\partial \tau_{xz}}{\partial x} + \frac{1}{\rho_m}\frac{\partial \tau_{yz}}{\partial y}$$

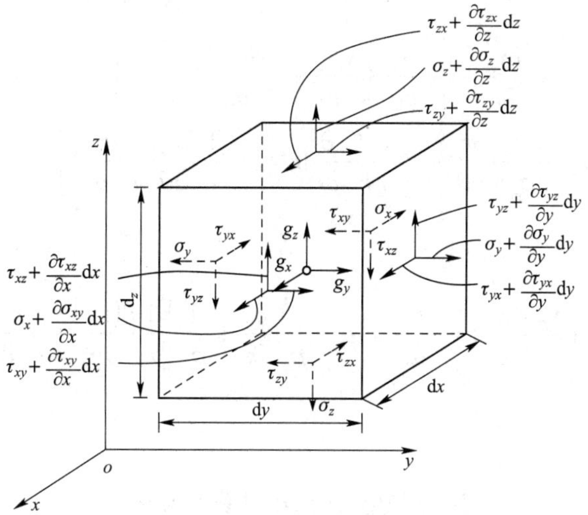

Figure 1.4 Surface stresses on a fluid element

These equations of motion are general without any restriction as to compressibility, viscous shear, turbulence, or other effects. The normal stresses can be rewritten as a function of the pressure p and the additional normal stresses, τ_{xx}, τ_{yy}, and τ_{zz}, accompanying deformation: $\sigma_x=-p+\tau_x$; $\sigma_y=-p+\tau_y$; $\sigma_z=-p+\tau_z$.

Cartesian coordinates:

x component

$$a_x = \frac{\partial v_x}{\partial t} + v_x\frac{\partial v_x}{\partial x} + v_y\frac{\partial y_x}{\partial y} + v_z\frac{\partial v_x}{\partial z} = g_x - \frac{1}{\rho_m}\frac{\partial p}{\partial x} + \frac{1}{\rho_m}\left(\frac{\partial \tau_{xx}}{\partial x} + \frac{\partial \tau_{yx}}{\partial y} + \frac{\partial \tau_{zx}}{\partial z}\right)$$

y component

$$a_y = \frac{\partial v_y}{\partial t} + v_x\frac{\partial v_y}{\partial x} + v_y\frac{\partial v_y}{\partial y} + v_z\frac{\partial v_y}{\partial z} = g_y - \frac{1}{\rho_m}\frac{\partial p}{\partial y} + \frac{1}{\rho_m}\left(\frac{\partial \tau_{xy}}{\partial x} + \frac{\partial \tau_{yy}}{\partial y} + \frac{\partial \tau_{zy}}{\partial z}\right) \tag{1.8}$$

z component

$$a_z = \frac{\partial v_z}{\partial t} + v_x\frac{\partial v_z}{\partial x} + v_y\frac{\partial v_z}{\partial y} + v_z\frac{\partial v_z}{\partial z} = g_z - \frac{1}{\rho_m}\frac{\partial p}{\partial z} + \frac{1}{\rho_m}\left(\frac{\partial \tau_{xz}}{\partial x} + \frac{\partial \tau_{yz}}{\partial y} + \frac{\partial \tau_{zz}}{\partial z}\right)$$

Cylindrical coordinates

r component

$$\frac{\partial v_r}{\partial t}+v_r\frac{\partial v_r}{\partial r}+\frac{v_\theta}{r}\frac{\partial v_r}{\partial \theta}-\frac{v_\theta^2}{r}+v_z\frac{\partial v_r}{\partial z}$$
$$=g_r-\frac{1}{\rho_{\mathrm{m}}}\frac{\partial p}{\partial r}+\frac{1}{\rho_{\mathrm{m}}}\left[\frac{1}{r}\frac{\partial}{\partial r}\left(rt_{rr}\right)+\frac{1}{r}\frac{\partial \tau_{\theta r}}{\partial \theta}-\frac{\tau_{\theta\theta}}{r}+\frac{\partial \tau_{zr}}{\partial z}\right]$$

θ Component

$$\frac{\partial v_\theta}{\partial t}+v_r\frac{\partial v_\theta}{\partial r}+\frac{v_\theta}{r}\frac{\partial v_\theta}{\partial \theta}+\frac{v_r v_\theta}{r}+v_z\frac{\partial v_\theta}{\partial z}$$
$$=g_\theta-\frac{1}{\rho_{\mathrm{m}} r}\frac{\partial p}{\partial \theta}+\frac{1}{\rho_{\mathrm{m}}}\left[\frac{1}{r^2}\frac{\partial}{\partial r}\left(r^2 t_{r\theta}\right)+\frac{1}{r}\frac{\partial \tau_{\theta\theta}}{\partial \theta}+\frac{\partial \tau_{z\theta}}{\partial z}\right] \tag{1.9}$$

z component

$$\frac{\partial v_z}{\partial t}+v_r\frac{\partial v_z}{\partial r}+\frac{v_\theta}{r}\frac{\partial v_z}{\partial \theta}+v_z\frac{\partial v_z}{\partial z}=g_z-\frac{1}{\rho_{\mathrm{m}}}\frac{\partial p}{\partial z}+\frac{1}{\rho_{\mathrm{m}}}\left[\frac{1}{r}\frac{\partial\left(\tau_{rz}\right)}{\partial r}+\frac{1}{r}\frac{\partial \tau_{\theta z}}{\partial \theta}+\frac{\partial \tau_{zz}}{\partial z}\right]$$

New Words and Phrase

1.kinematics *n.* 运动学

2.conversion *n.* 转换，变换

3.proportional *adj.* 成比例的

4.orthogonal *adj.* 正交的

5.magnitude *n.* 大小，量级

6.instant *n.* 瞬间

7.buoy *n.* 浮标

8.integral *n.* 积分

9.homogeneous *adj.* 均匀的

10.centroid *n.* （数）形心

Notes

1.As a natural science, the variability of river processes must be examined through the measurement of physical parameters.

就自然科学而言，河道的演变过程需要通过物理参数的测量来测定。

variability 可变性，变化性

With so much variability among these studies, it is possible that this review could significantly underestimate the number of patients with COPD who have concurrent anxiety, depression, or both.

2.The difference between the mass fluxes entering and leaving the differential control volume equal the rate of increase of internal mass.

流进和流出的质量通量的差异等于微分单元体的质量变化率。

differential 微分 mass flux 质量通量

Flood irrigation can be subdivided into basin irrigation, border irrigation and wild flooding.

3.The external forces per unit area applied on each face of the element are subdivided into normal-and tangential-stress components.

作用于单元体每个面上的外力可以分为法向和切向应力分量。

subdivide into　细分为

Flood irrigation can be subdivided into basin irrigation, border irrigation and wild flooding.

Comprehensive Exercises

I.Answer the following questions on the text.

1.What are the fundamental dimensions for the physical properties?

2.What is the relationship between shear stress τ_{zx} and dynamic viscosity μ?

3.What is the key for sediment transport problems?

4.What can the external forces applied on each face of the element be divided into?

5.What do the equations of motion in this passage neglect?

II.Fill the most appropriate words or phrases in the correct forms in the blanks from the list below.

measurement	throughout	prefer to	bring into	streamline
fill with	open channel	turbulence	designate	deformation

1.Some methods of_______are direct, but others are indirect.

2.Two types of conduits are used to convey water, the________and the pressure conduit.

3.Progressive abutment_______or yielding in response to arch thrust results in load-transfer and stress redistribution within the dam shell and in the abutment itself.

4.Clay (C)________for soil with combinations of PI and LL above the “A-line” for soils with PI> 7.

5.Settling tanks are designed to minimize________and allow the particles to fall to the bottom.

6.In the operation of reaction turbines, the runner chamber ________water and a draft tube is used to recover as much of the hydraulic head as possible.

7.With an axial-flow turbine, the generator may be placed in a________water-tight steel housing mounted in the center of the passageway.

8.When the inflowing sediment discharge exceeds the outgoing sediment capacity, alluvial channels tend to deposit their sediment load________the reach.

9.Some authorities_________classify these pumps separately from ordinary centrifugals because all impellers do not function entirely on the centrifugal principle.

10.During this period, the power station on the left bank and the permanent ship lock _____.

III.Translate the following sentences into Chinese from the text.

1.This passage first describes dimensions and units, physical properties of water. The equations governing the motion of water and sediment from upland areas to oceans include kinematics of flow, the equation of continuity, and the equation of motion.

2.The maximum water density is found at 4 °C, and water either colder or warmer than 4 °C will be found near the surface. The density of ice increases as the temperature decreases. This causes the ice cover to crack during cold nights and expand to apply large forces on the banks of lakes, reservoirs, and wide rivers during warm days.

3.For the particular case in which sediment diffusion is not significant, the conservation of solid mass is also defined after ρ_m is replaced with C_v. For sediment transport problems in which turbulent diffusion and dispersion are significant, sediment continuity equation, including turbulent-mixing coefficients, should be used.

4.The internal accelerations, or body forces per unit mass, acting at the center of mass of the element are denoted by g_x, g_y, and g_z. The external forces per unit area applied on each face of the element are subdivided into normal- and tangential-stress components. The normal stresses σ_x, σ_y, and σ_z are designated as positive for tension.

Reading Material Properties of sediment

The physical properties of sediment are classified into single particles, sediment mixture, and sediment suspension.

1. Single particle

The mass density of a solid particle, ρ_s, describes the solid mass per unit volume. The mass density of quartz particles, 2650 kg/m^3 (1 slug/ft, 3515.4 kg/m^3), does not vary significantly with temperature and is assumed constant in most calculations. It must be kept in mind, however, that heavy minerals such as iron, copper, etc., have much larger values of mass density.

Specific weight of solid particles, γ_s . The particle specific weight γ_s corresponds to the solid weight per unit volume of solid. Typical values of γ_s are 26.0 kN/m^3 or 165.4 lb/ft^3. The conversion factor is 1 lb/ft^3 = 157.09 N/m^3. The specific weight of a solid, γ_s, also equals the product of the mass density of a solid particle, ρ_s , times the gravitational acceleration g; thus

$$\gamma_s = \rho_s g \tag{1.10}$$

Specific gravity *G*. The ratio of the specific weight of a solid particle to the specific weight of fluid at a standard reference temperature defines the specific gravity *G*. With common reference to water at 4 ℃ , the specific gravity of quartz particles is

$$G = \frac{\gamma_s}{\gamma} = \frac{\rho_s}{\rho} = 2.65 \tag{1.11}$$

The specific gravity is a dimensionless ratio of specific weights, and thus its value remains independent of the system of units.

Submerged specific weight of a particle, γ_s. Owing to the Archimedes principle, the specific weight of a solid particle, γ_s, submerged in a fluid of specific weight γ equals the difference between the two specific weights; thus

$$\tilde{\gamma}_s = \gamma_s - \gamma = (G-1)\gamma \tag{1.12}$$

A wet-sieve method keeps the sieve screen and sand completely submerged. The sediment is washed onto the wet sieve and agitated somewhat vigorously in several directions until all particles smaller than the sieve openings have a chance to fall through the sieve. Material passing through the sieve with its wash water is then poured onto the next-smaller-size sieve. Particles retained on each sieve and those passing through the 0.062 mm sieve are transferred to containers that are suitable for drying the material and for obtaining the net weight of each fraction.

2. Sediment mixture

Particle-size distribution. The sediment size d_{50} for which 50% by weight of the material is finer is called the median grain size. Likewise d_{90} and d_{10} are values of grain size for which 90% and 10% of the material are finer, respectively.

Gradation coefficients σ_g and G_r. The gradation of the sediment mixture is a measure of non-uniformity of sediment mixtures. It can be described by

$$\sigma_g = \left(\frac{d_{84}}{d_{16}}\right)^{\frac{1}{2}} \tag{1.13}$$

or by the gradation coefficient

$$G_r = \frac{1}{2}\left(\frac{d_{84}}{d_{50}} + \frac{d_{50}}{d_{16}}\right) \tag{1.14}$$

Both gradation coefficients reduce to unity for uniform sediment mixtures, i.e., when $d_{84} = d_{50} = d_{16}$. The gradation coefficient increases with non-uniformity, and high-gradation coefficients describe well-graded mixtures.

Critical shear stress τ_c and shear velocity μ_{*c}. Approximate values of critical shear stress τ_c for non-cohesive particles can be obtained from the extended Shields diagram. The corresponding critical shear velocity μ_{*c} is defined as $u_{*c} = \sqrt{\tau_c / \rho}$. Note that both τ_c and μ_{*c} do not change significantly for sands and silts. To get crude approximations, a shear-stress value of $\tau =$ 0.1 Pa is sufficient to move silts but not sands, and $\tau_c =$ 1 Pa is sufficient to move sands but not gravels.

3. Sediment suspension

Volumetric sediment concentration C_v. The volumetric sediment concentration C_v is defined as the volume of solids $\forall_s$ over the total volume $\forall_t$. When the voids are completely filled with water, $\forall_v = \forall_w$, we obtain

$$C_v = \frac{\forall_s}{\forall_s + \forall_w} \tag{1.15}$$

The most common unit for sediment concentration is milligrams per liter, which describes the ratio of the mass of sediment particles to the volume of the water-sediment mixture. Other units include kilograms per cubic meter (1 mg/l = 1 g/m^3), the volumetric sediment concentration C_v, the concentration in parts in 10^6 (ppm) C_{ppm}, and the concentration by weight C_w. We can easily demonstrate the following identities:

$$C_w = \frac{\text{sediment weight}}{\text{total weight}} = \frac{C_v G}{1+(G-1)C_v} \tag{1.16}$$

in which $G = \gamma_s / \gamma$ and

$$C_{ppm} = 10^6 C_w \tag{1.17}$$

Note that the percentage by weight C_{ppm} is given by 1000000 times the weight of sediment over the weight of the water-sediment mixture. The corresponding concentration in milligrams per liter is then given by

$$C_{mg/l} = \frac{1mg/1GC_{ppm}}{G+(1-G)10^{-6}C_{ppm}} = \rho G C_v = \frac{10^6 mg}{1GC_v} \tag{1.18}$$

Settling velocity ω_0. The settling velocity ω_0 of sediment particles in clear water at 10 ℃ is calculated from

$$\omega_0 = \frac{8v}{d_s}\left\{\left[1+\frac{(G-1)g}{72v^2}d_s^3\right]^{0.5} - 1\right\} \tag{1.19}$$

where d_s is the particle diameter, v is the kinematic viscosity, G is the specific gravity, and g is the gravitational acceleration.

Specific weight of a mixture γ_m. The specific weight of a submerged mixture is the total weight of solid and water in the voids per unit total volume. The specific weight of a mixture, γ_m, is a function of the volumetric concentration C_v as

$$\gamma_m = \frac{\gamma_s \forall_s + \gamma \forall_v}{\forall_s + \forall_v} = \gamma_s (C_v) + \gamma(1 - C_v) \tag{1.20}$$

The specific mass ρ_m of a submerged mixture is the total mass of solid and water in the voids per unit total volume. The specific mass of a mixture is given by $\rho_m = \gamma_m / g$.

Porosity p_0. The porosity p_0 is a measure of the volume of void $\forall_v$ per total volume $\forall_t = \forall_v + \forall_s$.

The volume of solid particles $\forall_s = (1 - p_0)\forall_t$ is thus

$$p_0 = \frac{\forall_v}{\forall_t} = \frac{e}{1+e} \tag{1.21}$$

where the void ratio e is the ratio of the volume of void $\forall_v$ to the volume of solid $\forall_s$.

Dry specific weight of a mixture γ_{md}. The dry specific weight of a mixture is the weight of solid per unit total volume, including the volume of solids and voids. The dry specific weight of a mixture,γ_{md} is a function of porosity p_0 as

$$\gamma_{md} = \gamma_s(1 - p_0) = \gamma G(C_v) \tag{1.22}$$

Lesson 3 Soil Mechanics

Generally, most soil can be characterized as being made up of either or both of two distinctive types of grains. "Rounded" or "bulky" grains have a relatively small surface area with respect to their volume, similar to that of a sphere. These soil grains typically have little intragranular attraction (or bonds) and are therefore termed "cohesionless," referring to lack of tendency to "stick" together. Soil with these grain characteristics may also be called "granular." This soil group includes sands and gravels. Clay particles are very different, and are made of very thin plate-like grains, which generally have a very high surface to volume ratio. Because of this, the surface charges play a critical role in their intragranular attractive behavior and are termed "cohesive." This difference between grain types has a profound effect on behavior of a soil and the methodology by which improvement techniques can be effective.

1. Soil classification systems

There are a number of different soil classification systems that have been devised by various groups, which vary in definitions and categories of soil type. The Unified Soil Classification System (USCS; ASTM D2487) is dominant for most geotechnical engineers, as its soil type designations correlate well with many soils engineering properties. Thus, knowing a USCS designation may well be enough for a seasoned geotechnical engineer to be able to envision the types of properties such a soil may possess. The USCS will be used as the primary classification system throughout this text. Another common classification system, derived for use with roadway materials, is the American Association of State Highway and Transportation Officials (AASHTO) system (ASTM D3282, AASHTO M145). The AASHTO classification designations categorize soil types based on their usefulness in roadway construction applications. Another classification system is used by the US Department of Agriculture (USDA) for defining soil categories important for agricultural applications. The Massachusetts Institute of Technology also developed a soil classification system in which grain size definitions are nearly the same as the AASHTO. Table 1.1 and Figure 1.5 depict grain size definitions by various particle-size classification schemes. Soil classifications are typically limited to particle sizes less than about 76 mm (3 in).

Table 1.1 **Grain size definitions by various particle-size classification schemes**

Name of organization	Particle-size classifications Grain size (mm)			
	Gravel	Sand	Silt	Clay
Massachusetts Institute of Technology (MIT)	>2	2-0.06	0.06-0.002	<0.002
US Department of Agriculture (USDA)			0.05-0.002	<0.002
American Association of State Highway and Transportation Officials (AASHTO)	>2	2-0.05	0.075-0.002	<0.002
Unified Soil Classification System (US Army Corps of Engineers, US Bureau of Reclamation, and American Society for Testing and Materials)	76.2-2 76.2-4.75	2-0.075 4.75-0.075	Fines (i.e., silts and clays) <0.075	

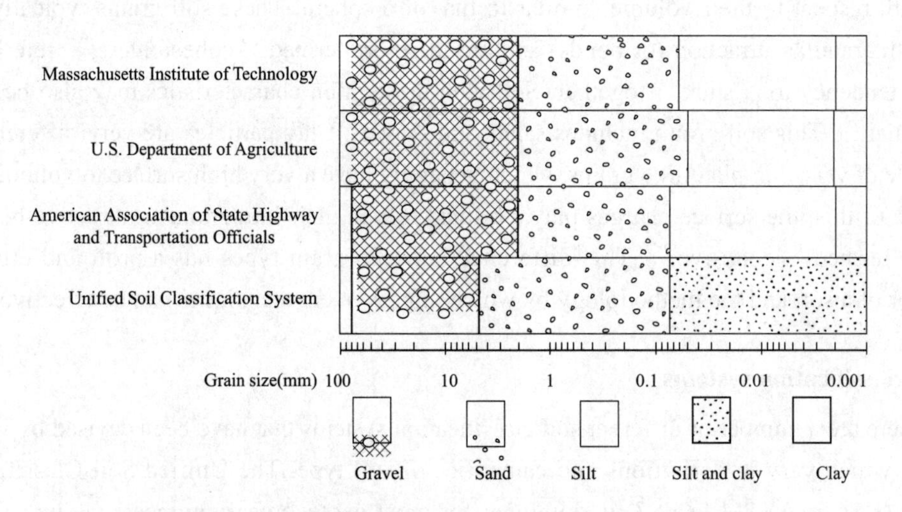

Figure 1.5 Grain size definitions by various particle-size classification schemes

2.Grain sizes and grain size distributions

At this point, one needs to clearly define a standard size to differentiate between coarse- and fine-grain sizes. This has been done for a number of classification systems using a standard screen mesh with 200 openings per inch, referred to as a No.200 sieve. The effective opening size of a No.200 sieve is 0.075 mm. Material able to pass through the No.200 sieve is termed "fine-grained" while that retained on the sieve is termed "coarse-grained". This standardized differentiation is not completely arbitrary or without merit as it is found that fine-grained soils tend to be more cohesive while coarse-grained soils are cohesionless. It is important to remember, however, that differentiation between clay and granular particles is not always represented by grain size and the No.200 sieve.

Analyzing the amounts or percentages of various grain size categories can be used to further classify soil types. Much can be ascertained by knowing the distribution of grain sizes, as these differences are related to various engineering properties and characteristics of soil. Common practice for coarse-grained soils is to filter a known amount (weight) of dry soil through a set of mesh screens or sieves with progressively smaller openings of known size. This will separate the soil into portions that pass one sieve size and are retained on another. This approach is known as a "sieve analysis" . Data of this type is collected such that the percentage passing each progressively smaller sieve opening size can be calculated. The results are presented as gradation plots or grain size distribution curves, plotted with percent passing versus nominal grain size. The grain size distribution is used for primary identification of coarse-grained soils and also can define gradation type.

Coarse-grained soils will generally fall into one of three different gradation types. Figure 1.6 depicts a representation of the general "shape" or trends of well-graded, poorly graded, and gap-graded soils. Well-graded soils span a wide range of grain sizes and include representation of percentages from intermediate sizes between the maximum and minimum sizes. Well-graded soils are often preferred as they are relatively easy to handle, can compact well, and often provide desirable engineering properties. Poorly graded (or well-sorted, or uniform) soils have a concentration of a limited range of grain sizes. This type of gradation can be found in nature due to natural phenomenon associated with depositional processes such as from alluvial and fluvial flows (rivers and deltas), waves (beach deposits), or wind (sand dunes). Poorly graded (or uniformly graded) soil gradations may be advantageous where seepage and ground water flow characteristics (drainage and filtering) are important. Uniformly graded soils can also be prepared manually by sieving techniques at small or large scale (such as for quarrying operations).

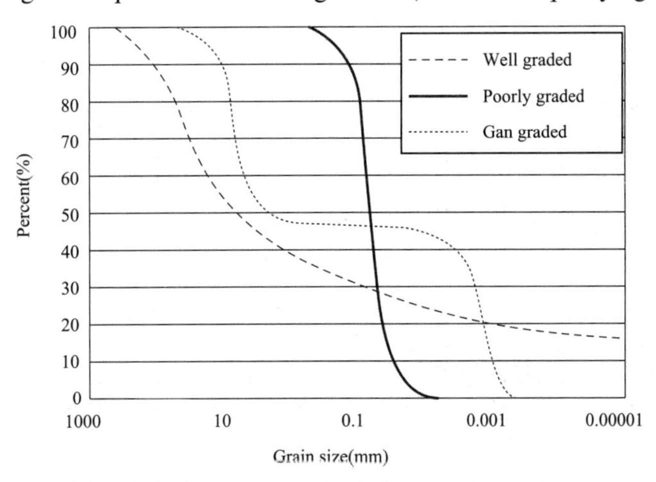

Figure 1.6 Representation of typical coarse soil gradation types

A third category for gradation is known as gap-graded, which refers to a soil with various grain sizes but which lacks representation of a range of intermediate sizes. Usually, this type of gradation is never desirable as it can create problems with handling and construction due to its tendency to segregate and create non-uniform fills. For classification purposes, gap-graded soils are considered to be a subset of poorly graded soils, as they are not well-graded.

3.Plasticity and soil structure

Classification schemes based solely on grain sizes (i.e., USDA) are relatively simple, but do not take into account the importance of clay properties on the behavioral characteristics of a soil. Both the USCS and AASHTO classification systems utilize a combination soil grain size distribution along with clay properties identifiable by plasticity of the finer-grained fraction of a soil. Plasticity is the ability of a soil to act in a plastic manner and is identified by a range of moisture contents where the soil is between a semisolid and viscous liquid form. These limits are determined as the plastic limit (PL) and liquid limit (LL) from simple, standardized laboratory index tests. For a more detailed discussion of these and related tests, refer to an introductory soil mechanics text, laboratory manual, or ASTM specifications (ASTM D4318).

Plasticity is commonly referred to by the Plasticity Index (PI), where PI¼ LL- PL. A graphical representation of plasticity developed for the purposes of classifying fine-grained soils gives the PI plotted as a function of LL (Fig. 1.7). The plot defines fine-grained soil classifications between clay and silt, and between high and low plasticity. There is a separating line called the A-line, defined by the equation PI¼ 0.73 (LL-20). Clay (C) is designated for soil with combinations of PI and LL above the "A-line" for soils with PI> 7. Soil below the A-line and PI> 4, and above the A-line with below PI< 4 are considered silt, designated "M". Another defining line is given for soils with LL above or below 50. Soils with LL> 50 are considered high plasticity, while those with LL< 50 are considered low plasticity. A special dual designation of CL-ML is given for soils above the A-line and 4 ≤ PI ≤ 7.

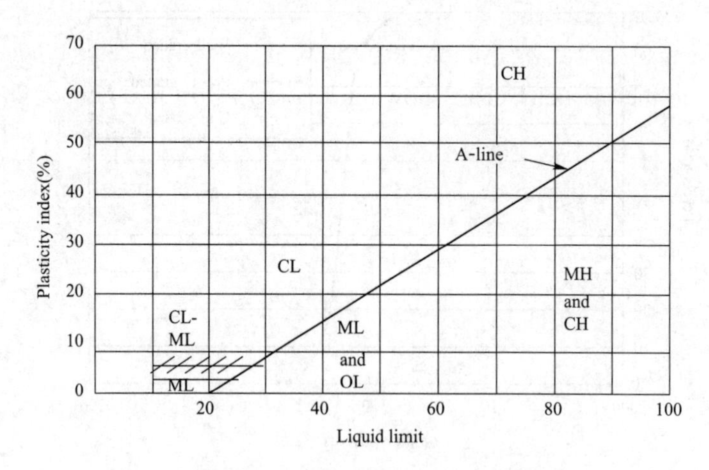

Figure 1.7 Plasticity chart for fine-grained soils

In order to classify a soil according to the USCS, a number of relatively simple steps must be followed. Only one to three simple index tests need to be performed in order to fully classify a soil: a sieve analysis, and/or a LL test, and a PL test. In the USCS, soil is generally classified by a two-letter designation. The first letter denotes the primary designation and identifies the dominant grain size or soil type. The primary designations are G, gravel; S, sand; M, silt; C, clay, O, organic, and Pt, peat (a highly organic soil). The second letter denotes a qualifier that provides further information regarding more detailed information on the makeup and characteristics of the soil.

Coarse-grained soils are defined as those where more than 50% of the soil is retained on the No. 200 sieve. According to the USCS, coarse soil grains retained on the No. 4 sieve (nominal opening size of 4.75 mm) are defined as gravel while those grains passing the No. 4 and retained on the No. 200 sieve are defined as sand. A coarse-grained soil is defined as gravel or sand depending on the dominant grain size percentage of the coarse fraction of the soil (where the coarse fraction is the cumulative percentage coarser than the No. 200 sieve). For example, if more than 50% of the material coarser than the No. 200 sieve is retained on the No. 4 sieve, then the soil is classified as gravel (G). If 50% or more of the material coarser than the No. 200 sieve passes the No. 4 sieve, then the soil is classified as sand (S).

For coarse-grained soils (G or S), the second qualifier denotes the type of gradation (P, poorly graded; W, well-graded) or the type of fine-grained soil contained if significant (M or C), so that coarse-grained soils will generally be classified with designations of GP, GW, GM, GC, SP, SW, SM, or SC. As mentioned earlier, fine-grained soils ("fines") become significant to the engineering properties and soil characteristics when as little as 5% by weight is contained. According to USCS, when less than 5% fine-grained material is present in a soil, fines are insignificant, and the second qualifier should pertain to the gradation characteristics according to the definitions provided below. The definition of well-graded versus poorly graded is a function of various grain sizes as determined by the grain size distributions. The definition of well-graded is based on two coefficients determined by grain sizes taken from the gradation curves.

Fine-grained soils (those where more than 50% of the soil passes the No.200 sieve) are defined according to the plasticity chart shown in Fig. 1.7. Most fine-grained soils will have a primary designation based on the LL versus PI values and their relationship to the "A-line" on the chart, with secondary designation as high (H) or low (L) plasticity, determined by whether the LL is above or below 50, respectively. Special cases for fine-grained soils are organic (O) designations OL and OH. Soils are determined to be organic based on changes in the LL as determined before and after oven drying. Other special cases of classification for fine-grained soils occur with low PI and LL values as seen on the plasticity chart (and described previously). AASHTO soil classification of fine-grained soils also uses a variation of a plasticity chart (see ASTM D3282).

Currently, ASTM D2487 utilizes the group symbol (two-letter designation) along with a group name, which can be determined using the same information gathered for classification designation, but adds a more detailed description that further elaborates on gradation. So for a complete classification and description including group name, one must know the percentages of gravel, sand and fines, and type of gradation (all based on sieve analyses), as well as LL and PI for fine-grained portions of the soil.

New Words and Phrase

1.grain *n.* 颗粒
2.stick *vi.* 黏
3.intragranular *adj.* 颗粒间
4.granular *adj.* 颗粒的
5.cohesive *adj.* 黏性的
6.geotechnical *adj.* 岩土的
7.opening *n.* 孔
8.grain size 粒径
9.fine-grained 细颗粒
10. arbitrary *adj.* 随意的，任意的
11. ascertain *vi.* 确定，查明
12. sieve analysis 筛分分析法
13. depositional *adj.* 淤积的
14. alluvial *adj.* 冲积的
15. quarry *vi.* 采石
16. segregate *vi.* 隔离，分离
17. plasticity *n.* 塑性，黏性
18. liquid limit 液限
19. dual *adj.* 双重的
20. airfield *n.* 飞机场
21. gravel *n.* 卵砾石
22. silt *n.* 粉沙
23. clay *n.* 黏土
24. peat *n.* 泥炭
25. qualifier *n.* 资格
26. nominal *adj.* 象征性的
27. pertain *vi.* 属于，适用于
28. elaborate *vi.* 详尽说明

Notes

1. "Rounded" or "bulky" grains have a relatively small surface area with respect to their volume, similar to that of a sphere.

圆形或者巨型颗粒在体积方面拥有相对较小的表面积，该情况与球体类似。

With respect to 关于，谈到 in respect of 关于，就……而言，比较

Heavy metals, organic pollutants, pathogenic organisms, etc., should all be understood with respect to their viability in natural systems and ability to cause environmental damage.

similar to 与……类似

Experiments similar to those reported here should be conducted using different age groups.

2.Classification schemes based solely on grain sizes (i.e., USDA) are relatively simple, but do not take into account the importance of clay properties on the behavioral characteristics of a soil.

仅根据颗粒粒径进行分类的方法相对简单，并未考虑黏性对土体力学特性的重要影响作用。

take into account 考虑，顾及

In deciding how to dispose of its wastes, an upstream firm or city is not forced to take into account the costs imposed by its effluent discharge upon downstream water users or the value of other uses of the water that may be foreclosed by its action.

3.A coarse-grained soil is defined as gravel or sand depending on the dominant grain size percentage of the coarse fraction of the soil (where the coarse fraction is the cumulative percentage coarser than the No. 200 sieve).

粗颗粒土是指级配曲线以卵砾石或者砂为主的土体（粗颗粒含量是指累积百分数大于第 200 号筛孔的占比）。

depend on 依靠、信赖，随……而定

The extent to which the measures can be applied will depend on the particular local circumstances. The GPS become not only accurate but (also) very fast measurement technique.

Comprehensive Exercises

I.Answer the following questions on the text.

1.What is the criteria to define the classification systems?

2.What is the grain size distribution curve?

3.Which has the lowest permeability, gravel, sand, silt, or clay?

4.What do take into account except the grain size in the USCS classification system?

II.Fill the most appropriate words or phrases in the correct forms in the blanks from the list below.

with respect to	granular	pass through	fall into	deposition
segregate	take into account	pertain	variation	elaborate

1.These particles are simply too large to ______the pore spaces of the medium and become trapped in the upper depths of the filter.

2._____soils consist predominantly of gravel, sand, and silt-sized particles.

3.The civil rights movement fought against practices that_____blacks and whites.

4.The location of stations______the line of protection should be selected for safe operation.

5.His creative output ______three distinct categories.

6.The sediment transport submodel______the influence of non-uniform sediment with bed surface armoring.

7.If accepted, mentors will be trained in all areas_______the program, and will receive ongoing support and training from the AHM Lanterns Mentoring Coordinator, Laurie Larsen.

8.He told the story in ______detail.

9.The physical properties of concrete are highly sensitive to_____in the mixture of the components.

10.He is quite interesting ______ the recent policies.

III.Translate the following sentences into Chinese from the text.

1.The phenomenon of the collapse of temporary, unconfined vertical surface comprised of granular soils is important for lightly loaded construction because it may make completing footing excavations and drilled shafts difficult owing to sloughing or collapsing of the sides.

2.Soil differs from most other engineering materials in that soil tends to fail in shear rather than a form of tension or compression.

3.The second letter denotes a qualifier that provides further information regarding more detailed information on the makeup and characteristics of the soil.

4.Most fine-grained soils will have a primary designation based on the LL versus PI values and their relationship to the "A-line" on the chart, with secondary designation as high (H) or low (L) plasticity, determined by whether the LL is above or below 50, respectively.

Reading Material Principal Design Parameters

In order to develop a plan of approach for designing a practical and economical solution, a geotechnical engineer must first initiate a stepwise process of identifying fundamental project parameters. These include: (1) establishing the scope of the problem, (2) investigating the conditions at the proposed site,(3) establishing a model for the subsurface to be analyzed, (4) determining required soil properties needed for analyses to evaluate engineering response characteristics, and (5) formulating a design to solve the problem. A number of engineering parameters that play critical roles in how the ground responds to various applications and loads typically need to be determined for each situation. Values of each parameter may be evaluated by field or laboratory tests of soils, or may be prescribed by design guidelines. Fundamental to applicable analyses and designs are input of reasonably accurate parameters that provide an estimate of response of the ground to expected loading conditions. Some of the parameters forming the basis of design applications are reviewed here.

Shear strength: Soil differs from most other engineering materials in that soil tends to fail in shear rather than a form of tension or compression. In fact, as soil exhibits very little tensile strength, convention is to take compression as positive and tension as negative, as opposed to standard mechanics of materials sign convention. Soil shear strength is then a function of the limiting shear stresses that may be induced without causing "failure". For the general case, shear strength is a function of frictional and cohesive parameters of a soil under given conditions of initial stresses and intergranular water pressures. Proper evaluation of shear strength is critical for many types of geotechnical designs and applications as it is fundamental to such considerations as bearing capacity (the ability for the ground to support load without failing), slope stability (an evaluation of the degree of safety for a soil slope to resist failure), durability (resistance to freeze-thaw and wet-dry cycles, as well as leaching for some soils), and liquefaction resistance.

Shear strength is a function of the effective confining stress (s_0). Here effective stresses are used as opposed to total stresses. Effective stresses are the intergranular stresses that remain after pore water pressures are accounted for. These are the actual stresses "felt" between grains, adding to their frictional resistance (strength). Total stresses are the combination of intergranular and pore water pressure acting on soil grains. Figure 1.8 graphically depicts shear strength as a function of effective confining stress in terms of a shear strength failure envelope. In looking at this figure, the plotted line defines the failure envelope. Any state of stress described by a point below the line is a possible state of equilibrium. Theoretically, once a state of stress is reached which touches the failure envelope, the soil will fail. Stress states above the failure envelope are not theoretically possible. Evaluation of the shear strength parameters $\acute{c}_0$ and F_0 may be obtained directly from laboratory tests or interpreted from in situ field tests performed as part of a site investigation.

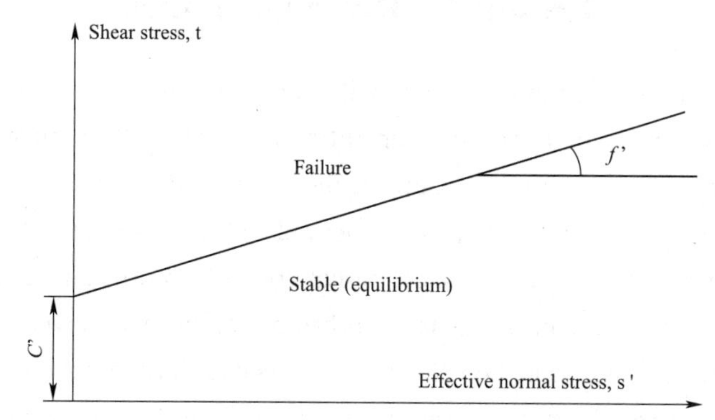

Figure 1.8 Graphical representation of the shear strength failure envelope

Laboratory tests typically used include: direct shear tests (ASTM D3080), unconfined compression tests (ASTM D2166), triaxial tests (ASTM D7181), and simple shear tests (ASTM D6528). In each of these tests (except unconfined compression), effective stress can be varied so that the shear strength (and shear strength parameters) can be evaluated for the appropriate stress levels estimated for each field application. The unconfined compression test may actually be considered a special case of the triaxial test, where the lateral confining stress is equal to zero. This test is simple and is often used as a quick indicator of strength and for comparative strength purposes, but is limited to cohesive soils (or in some cases, cemented soils).

A variety of in situ field tests are also available to evaluate soil shear strength. These include simple handheld devices such as the pocket penetrometer and pocket vane, which can give a quick estimate of strength for cohesive soils in a freshly excavated cut, trench, or pit. In situ tests such as the standard penetration test, vane shear, dilatometer, pressure meter, and shear wave velocity test can be performed in conventional boreholes as part of a field investigation.

The mechanism of bearing capacity failure is well documented and is described in detail in any text on shallow foundation design. While more detailed analyses address the finer aspects and contributions of irregular loads, footing shapes, slopes, and so forth, the fundamentals of foundation bearing capacity are dependent on size, shape, depth, and rigidity of a footing transmitting a level of applied stress to the supporting soil with respect to available resisting shear strength of the soil.

Slope stability may be simply described as the comparison of available resisting soil shear strength to the stresses applied by gravitational forces, and in more complicated situations, by water or seepage forces. Of course, there may be many more complexities involved, including geometry, soil variability, live or transient loads, dynamic loads, and so on, but in the context of soil improvement, any methods that increase the shear resistance of the soil along a potential shear surface beneath a slope will add to the stability. There are many applications of improvements and modifications that can solve a variety of slope stability issues.

Lesson 4　River Dynamics

Deviations from equilibrium conditions will trigger a dynamic response from the alluvial river system to restore the balance between inflowing and outflowing water and sediment discharges. Conceptually, the fluvial system of the watershed can be divided into three main zones: (1) an erosional zone of runoff production and sediment source; (2) a transport zone of water and sediment conveyance; and a depositional zone of runoff delivery and sedimentation. The second zone is characterized by near-equilibrium conditions between the inflow and the outflow of water and sediment. The bed elevation in this equilibrium zone is fairly constant. The upper zone is characterized by net erosion of bed material and channel degradation. The lower zone is characterized by net sedimentation and channel aggradation.

1.Riverbed degradation

Channel degradation refers to the general lowering of the bed elevation that is due to erosion. In some cases, the bed material is fine and degradation will result in channel incision. In other cases, the material is sufficiently coarse to form an armor layer that prevents further degradation.

1) Incised rivers

Slope adjustments refer to streams that would require either a steeper or a milder slope for reaching equilibrium between incoming and outgoing water and sediment discharges. Stated in simple terms, when the outgoing exceeds the in flowing sediment load, alluvial streams will scour bed material and degrade. Degradation results in channel incision and milder slopes.

Incised channels tend to be narrow and deep compared with equilibrium conditions. Channel incision will take place until equilibrium condition is reached. Incised channels are typical of upland areas whereby the sediment transport capacity increases in the downstream direction. Rills are small scale channels found in upland areas. Gullies are larger scale features also found in upland areas. Conventionally, rills can be crossed by farm machinery whereas gullies cannot.

In rivers, channel incision is found in arroyos and canyons. Arroyos are ephemeral channels in arid areas with flashy hydrographs that carry large sediment loads during short periods of time. Many arroyos dry out in the downstream direction as a result of infiltration and evaporation. The sediment load eventually deposits on the channel bed downstream of arroyos to form wide shallow streams. Canyons are usually deeply entrenched in vertical bedrock walls. Incised channels typically are narrower and deeper then equilibrium channels and are characterized by a shortage of sediment. Channel degradation also causes the banks to become unstable and subject to failure. Gully-like incised channels become very unstable, and bank erosion may become a significant source of sediment to the channel. Incised channels can often be found where the stream slope increases in the downstream direction. Knickpoints indicate points with a sudden change in bed slope. Headcuts usually refer to sudden drops in bed elevation. Headcuts usually start downstream, and their upstream migration is a characteristic feature of incised channels.

Degradation of the main river stem at river confluences causes headcutting and degradation in the tributaries. The headcut propagates upstream from the confluence and can cause severe stability problems in structures on shallow foundations such as bridges and some grade-control structures. Specific gauge records are often used to determine whether a stream tends to aggrade or degrade over time. A specific gauge record is the water-surface elevation that corresponds to a given discharge. When gauge records are available for a long period of time, plotting the gauge elevation at given discharges can detect long-term river trends. It is possible to observe opposite trends at different discharges; for instance, specific gauges may indicate a decreasing trend at a low discharge and an increasing trend at a high discharge. It is thus recommended to compare cross sections over time in order to confirm any trend detected with specific gauges.

2)Riverbed armoring

Armoring of the bed layer refers to coarsening of the bed material size as a result of degradation of well-graded sediment mixtures. The selective erosion of finer particles of the bed material leaves the coarser fractions of the mixture on the bed to induce coarsening of the bed material. When the applied bed shear stress is sufficiently large to mobilize the larger bed particles, degradation continues; when the applied bed shear stress cannot mobilize the coarse bed particles, an armor layer forms on the bed surface. The armor layer becomes coarser and thicker as the bed degrades until it is sufficiently thick to prevent any further degradation. The armor layer is representative of stable bed conditions and can be mobilized only during large floods. A riverbed is sometimes said to be paved when the armor layer can be mobilized only during exceptional floods. Three conditions need to be satisfied to form armor layers: (1) the stream must be degrading, (2) the bed material must be sufficiently coarse, and (3) there must be a sufficient quantity of coarse bed material. Relative to the first condition, the sediment transport capacity must exceed the sediment supply such that the stream attempts to scour the bed.

2.Riverbed aggradation

Channel aggradation refers to a gradual bed-elevation increase that is due to bedload sedimentation.

1)Braided rivers

When the inflowing sediment discharge exceeds the outgoing sediment capacity, alluvial channels tend to deposit their sediment load throughout the reach. Streams carrying mostly wash load will not change their morphology because the sediment overload will be carried downstream to settle in lakes, reservoirs, or estuaries. Streams carrying most of their sediment load in suspension change their morphology gradually as the excess sediment load settles in the down-stream direction. The riverbed material size becomes gradually finer in the downstream direction. From Lane's relationship, downstream fining is usually accompanied by a downstream decrease in bed slope. On the other hand, streams that carry predominantly bed load material will respond quite rapidly to a change in sediment-transport capacity.

A decrease in transport capacity induces direct settling on the bed of alluvial channels. As sketched in Fig. 1.9, the settling of bed load forces aggrading channels out of the bankfull conditions. The flow spreads on the floodplain with accumulation of the bed-sediment load to form natural levees on a wide floodplain. There is a tendency for the stream to widen and become very shallow with bars subjected to rapid changes in morphology. At high flows, braided streams have a low sinuosity and often appear to be straight. At low flows, numerous small channels weave through the exposed bars. These streams are known to braid as the bed slope increases through aggradation. The flow velocity of braided streams is high, and the bed material can be easily mobilized. Braided streams are rather unstable in that they are prone to severe lateral migration, frequent shifts, and changes in cross-section geometry. The bars of braided streams are generally submerged once a year and are devoid of vegetation.

Islands are different from bars in that they are stabilized by vegetation and rivers with multiple islands are anastomosed. Anastomosed rivers are usually more stable than braided channels because vegetation straightens the banks and stable islands control the flow between the branches. During floods, vegetated islands trap sediment and aggrade. Because braided channels require large bed load transport, most braided rivers are steep, and therefore, at a given discharge, braided rivers should be steeper than meandering rivers. The range of slope variability, however, is quite extensive, and it remains difficult to separate braiding from meandering channels solely on the basis of bankfull discharge and slope. An alternative approach is based on the width–depth ratio. In general, braided channels have a width-depth ratio in excess of 100. Several criteria based on bankfull discharge and slope: a single channel with given dominant discharge is thus thought to meander on mild slopes and braid on steep slopes. This concept has been expanded by Lane (1957) who proposed a slope-discharge relation for sand-bed channels. Empirically, braided channels were observed when SQ1/4 > 0.01 and channels.

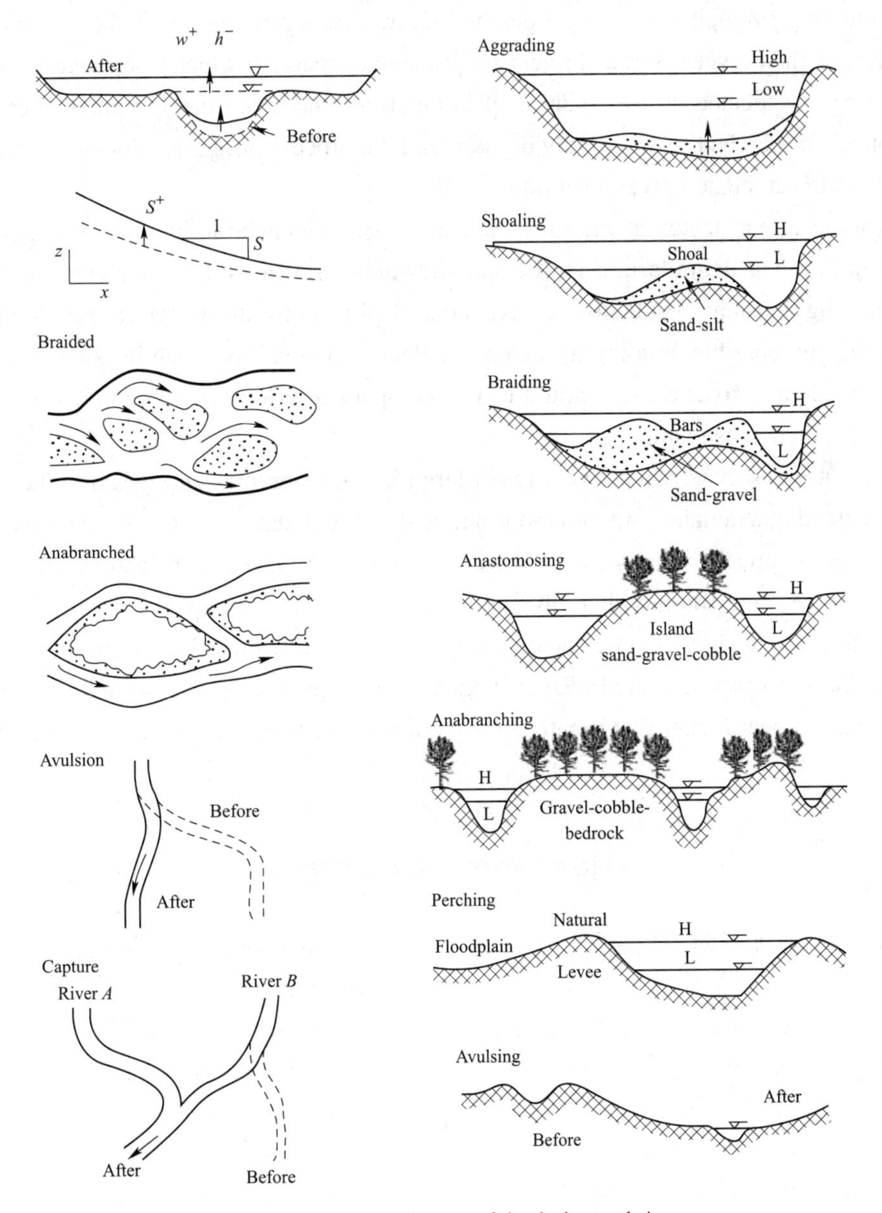

Figure 1.9 Schematic features of riverbed aggradation

2)Alluvial fans and deltas

Alluvial fans are found where steep mountain channels reach valley floors. The sudden break in bed slope causes the bed material transported by the river to deposit. The accumulation of debris usually takes a conical shape. The volume of material in the alluvial fan is indicative of the sediment transport capacity of the stream through geologic times. The aggradation takes place on the riverbed and on natural levees between the apex of the alluvial fan and the valley floor.

Through aggradation and natural levee formation, a river raises its own bed elevation above the surrounding floodplain to form a perched river. Perched rivers are stable as long as they cannot breach their levees. Perched rivers are prone to avulsion in which rivers select a new flow path that can be located up to hundreds of kilometers away from their original river courses. Old channels of perched rivers rapidly dry out, and the process of aggradation and natural levee formation starts at the new river location.

To some extent, tectonic activities exert a similar influence on alluvial river morphology as aggradation and degradation. Uplift and subsidence on the side of a river may result in river perching and may result in river avulsions. Uplift along the water course should cause aggradation and possible braiding upstream and degradation and possible incision downstream. Subsidence along a river course should cause the opposite effects, with degradation upstream and aggradation downstream.

Deltas are observed when rivers reach large lakes, reservoirs and oceans. The sediment deposits extend in a deltaic form into the water, and the aggradation in the upstream river causes the deposits to spread laterally in the lower reaches of the river. The features of deltas and alluvial fans are quite similar where the valley floor is replaced with the water level and deltas are much flatter than alluvial fans. The delta deposits are usually fine (finer than sand) as opposed to alluvial fans (coarser than sand). Rivers transporting large quantities of wash load may remain sinuous and meander to the river mouth. Rivers transporting large quantities of bed material tend to braid.

New Words and Phrase

1.deviation *n.* 偏离；偏差
2.fluvial *adj.* 河流的
3.conveyance *n.* 传送；运送；输送
4.degradation *n.* 恶化；堕落
5.incision *n.* 下切
6.armor *vt.* 为……装甲
7.mild *adj.* 温和的，轻微的
8.rill *n.* 小溪，小河
9.gully *n.* 沟道
10.arroyo *n.* 小河，河谷
11.canyon *n.* 峡谷
12.ephemeral *adj.* 短暂的
13.hydrograph *n.* 水位图
14.infiltration *n.* 渗透
15.kinckpoint *n.* 尼克点
16.confluence *n.* 汇合口
17.tributary *n.* 支流
18.mobilize *vt.* 使……流通，松动
19.aggradation *n.* 沉积
20.braided river 辫状河道
21.reservoir *n.* 水库
22.bankfull *n.* 平摊
23.floodplain *n.* 洪泛平源
24.devoid *adj.* 缺乏的
25.anastomosed river 网状河道
26.conical *adj.* 圆锥形的
27.tectonic *adj.* 构造的
28.subsidence *n.* 下沉

29.sinuous *adj*. 弯曲的，蜿蜒的

Notes

1.Deviations from equilibrium conditions will trigger a dynamic response from the alluvial river system to restore the balance between inflowing and outflowing water and sediment discharges.

冲积河流系统相应于平衡状态的偏离将促发动态调整来恢复进出水沙量的平衡。

trigger 引发，引起；触发

A shortfall in the availability of water which prevents the attainment of planned production levels, which are otherwise capable of being achieved, will be treated as an event of Force Majeure and will trigger the release of a subordinated loan from the ESF.

restore 恢复；修复 alluvial river system 冲积河流系统

2.In some cases, the bed material is fine and degradation will result in channel incision. In other cases, the material is sufficiently coarse to form an armor layer that prevents further degradation.

一些情况下，若床沙较细将会引起河床退化，从而导致河床下切。另外一些情况下，粒径粗的床沙将会在床面上形成粗化层，从而保护河床高程的进一步下降。

result in 引起，后接结果 armor layer 粗化层

It is also noted that a future increase in building loads due to a new addition, for example, may result in additional immediate settlements.

3.To some extent, tectonic activities exert a similar influence on alluvial river morphology as aggradation and degradation.

某种程度上，构造运动对冲积河流地貌的抬升和下降起到了类似的影响。

to some extent 某种程度上 in addition to 运用；行使；施加

They are the basis for many plastics and are now being used to produce concrete with more strength and durability. Accordingly, they help cut concrete weight to some extent.

Comprehensive Exercises

I.Answer the following questions on the text.

1.How many zones can the watershed be divided into, what are they?

2.What is the channel degradation?

3.Where can we find the channel incision in the Earth?

4.What type of channel pattern forms when the inflowing sediment discharge exceeds the outgoing sediment capacity?

5.What the characteristics of the alluvial deltas?

II.Fill the most appropriate words or phrases in the correct forms in the blanks from the list below.

conveyance	due to	armor layer	infiltration	propagate
correspond to	relative to	deposit	avulsion	subsidence

1.Surface water flow onto golf courses can change over time _____local developments and new roads, causing more water to drain onto the property.

2.The prestressed steel roads are placed in the lower part of a form that ______the shape of the finished structure, and the concrete is poured around them.

3.The capacity of canals shall be sufficient to meet the amount of water needed to cover the estimated______losses in the canal.

4.When the applied bed shear stress cannot mobilize the coarse bed particles,_____forms on the bed surface.

5.The silt which_____in the yearly floods, and made the Nile floodplain fertile, is now held behind the dam.

6.Coastal lands, such as offshore bars formed by sediments carried by coastal currents occupy a position _____ the sea that floodplains do to rivers.

7.Water moves into soil by _____at low rates of less than 2.5 mm per hour to high rates of over 25 mm per hour.

8.The easiest way to _____a vine is to take hardwood cuttings.

9.In sedimentary geology and fluvial geomorphology,______is the rapid abandonment of a river channel.

10.Many basins were formed by the _____of the Earth's crust.

III.Translate the following sentences into Chinese from the text.

1.Incised channels tend to be narrow and deep compared with equilibrium conditions. Channel incision will take place until equilibrium condition is reached. Incised channels are typical of upland areas whereby the sediment transport capacity increases in the downstream direction.

2.Three conditions need to be satisfied to form armor layers: (1) the stream must be degrading, (2) the bed material must be sufficiently coarse, and (3) there must be a sufficient quantity of coarse bed material. Relative to the first condition, the sediment transport capacity must exceed the sediment supply such that the stream attempts to scour the bed.

3.These streams are known to braid as the bed slope increases through aggradation. The flow velocity of braided streams is high, and the bed material can be easily mobilized. Braided streams are rather unstable in that they are prone to severe lateral migration, frequent shifts, and changes in cross-section geometry.

4.The features of deltas and alluvial fans are quite similar where the valley floor is replaced with the water level and deltas are much flatter than alluvial fans. The delta deposits are usually fine (finer than sand) as opposed to alluvial fans (coarser than sand).

专业术语释义与科技英语写法

1. 基础理论其他主要术语释义

1.1 水流基本词汇

flow discharge 流量
turbulent flow 紊流
laminar flow 层流
unsteady flow 非恒定流
uniform flow 均匀流
supercritical flow 急流
local head loss 局部水头损失
uniform seepage flow 均匀渗流
hydrostatic force 静水压力
continuous equation 连续方程
energy equation 能量方程
critical depth 临界水深
stream line 流线
open channel 明渠
shear stress 切应力
viscous force 黏滞力
potential flow 势流
hydraulic radius 水力半径
water head 水头
water level 水位
hydraulich jump 水跃
frictional head loss 沿程水头损失
weir flow 堰流
backwater curve 壅水线
sluice flow 闸口出流
gross flow 总流
gross head 总水头
Fround number 弗劳德数
Reynolds number 雷诺数
Chezy formula 谢才公式

1.2 泥沙基本词汇

sediment 泥沙
nominal diameter 等容粒径
sieve diameter 筛分粒径
cobble 卵石
boulder 漂石
specific weight 干重度
angle of repose 水下休止角
median size 中值粒径
fall velocity 沉速
bed form 床面形态
ripple 沙纹
dune 沙垄
sand wave 沙波
bed load 推移质
suspended load 悬移质
bed material load 床沙质
wash load 冲泻质
sediment transport capacity 输沙率
meandering river 弯曲河流
braided river 辫状河流
straight river 顺直河流
wandering river 游荡型河流

2. 科技英语特点

科技英语由于其内容、使用域和语篇功能的特殊性，也由于科技工作者长期以来的语言使用习惯，形成了自身的一些特点，使其在许多方面有别于日常英语、文学英语等语体。科技文章具有准确、客观、正式、逻辑严密等特点，主要体现在以下几方面：

1）大量使用名词化结构

《当代英语语法》（A Grammar of Contemporary）在论述科技英语时提出，大量使用名词化结构（Nominalization）是科技英语的特点之一。因为科技文体要求行文简洁、表达客观、内容确切、信息量大，强调存在的事实，而非某一行为。例如：Archimedes first discovered the principle of displacement of water by solid bodies.（阿基米德最先发展固体排水的原理。）句中 of displacement of water by solid bodies 系名词化结构，一方面简化了同位语从句，另一方面强调了 displacement 这一事实。

2）广泛使用被动语句

根据英国利兹大学 John Swales 的统计，科技英语中的谓语至少 1/3 是被动态。这是因为科技文章侧重叙事推理，强调客观准确，因此采用被动语态，例如：Attention must be paid to the working temperature of the machine.（应当注意机器的工作温度。）

3）第三人称句多

科技英语文体的一个显著特点就是很少有第一、第二人称句，这是由于科技文体的主要目的在于阐述科学事实、科学发现、实验结果等。第一、二人称使用过多，会造成主观臆断的印象，因此尽量使用第三人称叙述。

4）非限定动词的应用和大量使用后置定语

科技文章要求行文简练、结构紧凑，往往使用分词短语代替定语从句或状语从句；使用分词独立结构代替状语从句或并列分句；使用不定式短语代替各种从句；介词 + 动名词短语代替定语从句或状语从句。可缩短句子，又比较醒目。

5）大量使用常用句型

科技文章中经常使用若干特定的句型，从而形成科技文体区别于其他文体的标志。例如 It…that 结构句型；被动态结构句型；结构句型，分词短语结构句型，省略句结构句型等。举例如下：It is evident that a well lubricated bearing turns more easily than a dry one.（显然，润滑好的轴承，比不润滑的轴承容易转动。）

6）为了描述精确，使用长句多

为了表述一个复杂概念，使之逻辑严密，结构紧凑，科技文章中往往出现许多长句。有的长句多达七八个词，例句省略。

7）大量使用复合词与缩略词

大量使用复合词与缩略词是科技文章的特点之一，复合词从过去的双词组合发展到多词组合；缩略词趋向于任意构词，例如某一篇论文的作者可以就仅在该文中使用的术语组成缩略词，这给翻译工作带来一定的困难。例如：full-enclosed 全封闭的（双词合成形容词）。

3. 科技英语翻译的标准

关于翻译的标准，历来提法很多。清末启蒙思想家严复主张“信、达、雅”，当代翻译界主张“忠实、通顺”。就科技英语的特点和用途而言，其翻译的标准应略有区别于文学翻译。科技英语翻译主要是两种语言间的信息转换，为此，笔者认为，在进行科技英语翻译时要坚持三条标准：①准确规范。所谓准确就是忠实地、不折不扣地传达原文的全部信息内容；所谓规范就是译文要符合所涉及的科学技术或某个专业领域的专业语言表达规范。②通顺易懂。译文语言必须通顺，符合规范，用词造句应符合本民族语言的习惯，要用民族的、科学的、大众的语言，以求通顺易懂。不应有文理不通、逐词死译和生硬晦涩等现象。③简洁明晰。就是译文要简短精练、一目了然，要尽量避免烦琐、冗赘和不必要的重复。

4. 科技英语的翻译技巧

要使译文达到“准确规范”“通顺易懂”“简洁明晰”这三个标准，就必须运用翻译技巧。翻译技巧就是在翻译过程中用词造句的处理方法，如词义的引申、增减、词类转换和科技术语的翻译方法等。

1）引申译法

当英语句子中的某个词按词典的释义直译不符合汉语修辞习惯或语言规范时，则可以在不脱离该英语词本义的前提下，灵活选择恰当的汉语词语或词组译出。例如：Jigang will fix this problem during the recent shut down of the finishing mill.（济钢会在最近的精轧机停产时解决这一问题。）“fix”字典意思为“固定、修理”，这里引申译为“解决、处理”。

2）增减词译法

增词就是在译句中增加或补充英语句子中原来没有或省略了的词语，以便更完善、更清楚地表达英语句子所阐述的内容。在英语句子中，有的词从语法结构上讲是必不可少的，但并无什么实际意义，只是在句子中起着单纯的语法作用；有的词虽有实际意义，但按照字面译出又显多余。这样的词在翻译时往往可以省略不译。

3）词类转换

英语翻译中，常常需要将英语句子中属于某种词类的词，译成另一种词类的汉语词，以适应汉语的表达习惯或达到某种修辞目的。这种翻译处理方法就是转换词性法，简称词类转换。例如：In any case, the performance test has priority.（不管怎样进行，性能测试都要优先。）这里将名词“priority”转译为动词“优先”。

4）词序处理法

英汉两种语言的词序规则基本相同，但也存在着某些差别。不同的英语句子，在翻译中的词序处理方式也常常不同。例如：An insufficient power supply makes the motor immovable.（电源不足就会使马达停转。）这里将“insufficient power”（不足电源）改序翻译为“电源不足”较为合理。

5）科技术语的译法

工程英语中有大量的科技术语，而且科学性、技术性和专业性很强。科技术语的译法有意译、音译、象形译和原形译 4 种。例如 Final as built drawings（竣工图），first aid box（急救箱），centrifugal force（离心力），gear box（齿轮箱、变速箱）。

UNIT II   WATER WORK

Lesson 1   Reservoirs and Dams

Dams and reservoirs are introduced because they provide high-level controls on natural water systems for all water uses, including water supply, hydropower, navigation, fish and wildlife, and recreation. They also influence stream water quality and have a central role in flood control.

1. Dams

Dam are water-retaining structures which, effecting a closure of the stream, create heading-up. It is a structure built across a stream, river, or estuary to retain water. Its purposes are to meet demands for water for human consumption, irrigation, or industry; to reduce peak discharge of flood water; to increase available water stored for generating hydroelectric power; or to increase the depth of water in a river so as to improve navigation. An incidental purpose can be to provide a lake for recreation.

Dams are classified on the basis of the type and materials of construction, as gravity, arch, buttress, and earth. The first three types are usually constructed of concrete. A gravity dam depends on its own weight for stability and is usually straight in plan although sometimes slightly curved. Arch dams transmit most of the horizontal thrust of the water behind them to the abutments by arch action and have thinner cross sections than comparable gravity dams. Arch dams can be used only in narrow canyons where the walls are capable of withstanding the thrust produced by the arch action. The simplest of the many types of buttress dams is the slab type, which consists of sloping flat slabs supported an intervals by buttresses. Earth dams are embankments of rock or earth with provision for controlling seepage by means of an impermeable core or upstream blanket. More than one type of dam may be included in a single structure. Long dams often have a concrete river section containing spillway and sluice gates and earth or rock-fill wing dams for the remainder of their length.

1) Gravity dams

Gravity dams refer to solid concrete or masonry dams of roughly triangular cross section, which depend primarily on their own weight and cohesion with the foundation for stability. The dams are usually straight, but may be slightly curved in plan. Gravity dams may be triangular in cross section with the bed width almost 2/3 of its height, or trapezoidal. In spite of their impressive bulk, gravity dams usually require the same factor of safety as thin arch dams (Fig. 2.1).

Figure 2.1 The typical gravity dam

2) Arch dams

Generally, the arch dams are used in narrow rocky canyons. Arch dams transfer the greater proportion of the water load to the valley sides rather than to the floor. Abutment integrity and stability are therefore critical, and the importance of this point cannot be overstated. Progressive abutment deformation or yielding in response to arch thrust results in load-transfer and stress redistribution within the dam shell and in the abutment itself. In more extreme situations of significant abutment yielding or instability local overstress of the dam wall will result in catastrophic collapse（Fig. 2.2）.

Figure 2.2 The typical arch dam

3) Earth fill dams

An earth fill dam can provide a cost-effective method of storing larger volumes of water. The earth fill dams can be defined as a dam constructed from natural materials excavated or obtained nearby. The nature fill materials are placed and compacted without the addition of any binding agent, using high-capacity mechanical plant. In order to be effective, a dam wall must remain stable during large storm events and the soil must be relatively impermeable to minimize seepage loss（Fig. 2.3）.

Figure 2.3 The typical earth fill dam

2.Reservoir

A reservoir has unique operational and maintenance characteristics compared to natural lakes. Unlike lakes, reservoirs may have extreme water level fluctuations and therefore a high potential for shoreline erosion and instability. The water stored in reservoir may be used for various purposes. Depending upon the purposes served, the reservoir may be classified as follows: 1) storage or conservation reservoir; 2) flood control reservoirs; 3) distribution reservoirs; 4) multipurpose reservoir.

1) Storage or conservation reservoirs

A city water supply, irrigation water supply or a hydroelectric project drawing water directly from a river or a stream may fail to satisfy the consumers (consumers') demands during extremely low flows, while during high flows, it may become difficult to carry out their operation hydroelectric project due to devastating floods. A storage or a conservation reservoir can retain such excess supplies during periods of peak flows and can release them gradually during low flows as and when the need arises.

A storage reservoir with gated spillway and gated sluiceway, provides more flexibility of operation, and thus gives us better control and increased usefulness of the reservoir. Storage reservoir are, therefore, preferred on large rivers which require better control, while retarding basins are preferred on small rivers. In storage reservoirs, the flood crest downstream can be better controlled and regulated properly so as not to cause their coincidence. This is the biggest advantage of such a reservoir and outweighs its disadvantages of being costly and involving risk of human error in installation and operation of gates. Incidentally, in addition to conserving water for later use, the storage of floodwaters may also reduce flood damage below the reservoir. Hence, a reservoir can be used for controlling floods either solely or in addition to other purposes.

2) Flood control reservoir

A flood control reservoir or generally called flood-mitigation reservoir, stores a portion of the flood flows in such a way as to minimize the flood peaks at the areas to be protected downstream. To accomplish this, the entire inflow entering the reservoir is discharged till the outflow reaches the safe capacity of the channel downstream. The inflow in excess of this rate is stored in the reservoir, which is then gradually released so as to recover the storage capacity for next flood. A reservoir with gates and valves installation at the spillway and at the sluice outlets known as a storage-reservoir, while on the other hand, a reservoir with fixed ungated outlets is known as a retarding basin.

A retarding basin is usually provided with an uncontrolled spillway and an uncontrolled orifice type sluiceway. The automatic regulation of outflow depending upon the availability of water, takes place from such a reservoir. The maximum discharging capacity of such a reservoir should be equal to maximum safe carrying capacity of the channel downstream. As flood occurs, the reservoir gets filled and discharges through sluiceways. As the reservoir elevation increases, outflow discharge increases. The water level goes on rising until the flood has subsided and the inflow becomes equal to or less than the outflow. After this, water gets automatically withdrawn from the reservoir until the stored water is completely discharged.

The advantages of a retarding basin over a gate controlled detention basin are: ① cost of gate installations is saved. ② there are no gates and hence, the possibility of human error and negligence in their operation is eliminated. ③ since such a reservoir is not always filled, much of land below the maximum reservoir level will be submerged only temporarily and occasionally and can be used for agriculture, although no permanent habitation can be allowed on this land.

3) Multipurpose reservoirs

A reservoir planned and constructed to serve not only one purpose but various purposes together is called a multipurpose reservoir. Reservoir, designed for one purpose, incidentally serving other purposes, shall not be called a multipurpose reservoir, but will be called so, only if designed to serve those purposes also in addition to its main purpose. These reservoirs store water during runoff periods for use during other times of the year and provide a degree of flood control. Reservoir water is used to supply the needs of municipalities, industrial users, hydropower generation, and the farming community. It is also used to provide support to the aquatic and riparian environment downstream of the reservoir.

4) Distribution reservoir

A distribution reservoir is a small storage reservoir constructed within a city water supply system. Such a reservoir can be filled by pumping water at a certain rate and can be used to supply water even at rates higher than the inflow rate during periods of maximum demands (called critical periods of demand). Such reservoirs are, therefore, helpful in permitting the pumps or water treatment plants to work at a uniform rate, and they store water during the hours of no demand or less demand and supply water from their "storage" during the critical periods of maximum demand.

New Words and Phrase

1.afflux *n.* 流入
2.gravity *n.* 重力
3.triangular *adj.* 三角的
4.trapezoidal *adj.* 梯形的
5.buttress *n.* 扶壁，支墩
6.cross section 断面
7.seepage *n.* 渗漏
8.earth dam 土坝
9.withstand *vt.* 经受，承受
10.slab *n.* 平板
11.spillway *n.* 溢洪道
12.binding agent 黏合剂
13.compact *vi.* 压紧，压实
14.potential *n.* 势能
15.sluiceway *n.* 闸门
16.retarding basin 滞洪区
17.flood crest 洪峰
18.valve *n.* 闸
19.outlet *n.* 出口
20.orifice *n.* 孔口
21.hydropower generation 水力发电
22.riparian *adj.* 河边的
23.distribution reservoir 配水水库
24.multipurpose reservoir 综合水库
25.flood control reservoir 防洪水库
26.storage reservoir 蓄水水库

Notes

1.Progressive abutment deformation or yielding in response to arch thrust results in load-transfer and stress redistribution within the dam shell and in the abutment itself.

拱推力形成的渐进的变形或者屈服将引起坝壳和支座的荷载传送和应力重分布。

in response to 对……做出反应

Biden's statement in response to Flores's and Lappos's allegations seems to attempt to clarify intent.

result in 导致，引起，后面通常接结果

Floods often result in a loss of life especially in low-lying areas and along river banks.

2.The inflow in excess of this rate is stored in the reservoir, which is then gradually released so as to recover the storage capacity for next flood.

超过入库率的水体将被存储在水库中，随后将逐渐释放，以便为下次洪水提供存储库容。

so as to 为了；以便；以致；为的是 storage capacity 存储库容

3.Reservoir, designed for one purpose, incidentally serving other purposes, shall not be called a multipurpose reservoir, but will be called so, only if designed to serve those purposes also in addition to its main purpose.

水库设计阶段是服务一个目标，随带可为其他目标服务的不能称为综合型水库，仅有当其在设计阶段考虑主要服务目标的同时兼具其他目标，才能称之为综合型水库。

in addition to　除……之外

In addition to the direct use of water in our homes and on the farm, there are many indirect ways in which water affects our lives.

Comprehensive Exercises

I.Answer the following questions on the text.

1.What are the materials used to construct dams?

2.How does an arch dam counteract the force of water pressure?

3.Why do the gravity dam have a triangular profile?

4.What are the functions of the gates installed in a reservoir?

5.How are reservoirs classified according to the text, and what are they?

II.Fill the most appropriate words or phrases in the correct forms in the blanks from the list below.

navigation	retain	by means of	spillway	rather than
capacity	take place	withdrawn	potential	redistribution

1.______conditions both upstream and downstream on the Yangtze River have been improved in the low flow season.

2.The turbine blades can be made to run ______flowing water.

3.______are the hydraulic passageway designed to conduct flood flows safely past the dam.

4.The prediction relationship for ______reservoir systems is important because it can be used as an alternative to the analysis of reservoirs during designing stage.

5.The basic functions of dams, to control stream flows and impound water, create many situations, as stream flows ______from the smallest headwaters to giant bays and estuaries that empty into the ocean.

6.A dam normally has a service______for ongoing releases and an emergency spillway to release large storm flows.

7.Slow earthquakes provide a mechanism for stress______ before normal earthquakes.

8.After the war, the construction of the bridges______.

9.Reservoirs may have extreme water level fluctuations and therefor a high ______for bank stability.

10.Systems in this landscape are designed more for waterlogging and inundation management ______ salinity control.

III.Translate the following sentences into Chinese from the text.

1.Dams are water-retaining structures which, effecting a closure of the stream, create heading-up. Dams is made of various materials, such as concrete, reinforced- concrete, stone and earth, respectively. Concrete and embankment dams are most common, the latter being constructed of various ground materials including sand, loamy sand, loams, gravel, pebbles, combination of these, and of quarry stone.

2.A reservoir is a lake where water is stored naturally or artificially and a pond is a small reservoir. In the United States, there are countless thousands of small lakes, compared to relatively few giant reservoirs. These small impoundments even extend down to the level of urban storm water detention ponds, which function the same way as flood control reservoirs, but hold less water and respond faster to inflows.

3.A concrete gravity dam is normally a massive structure with a pervious foundation, a cutoff wall, a downstream and upstream apron, and anchor walls to aid in prevention of sliding.

4.Such reservoirs are, therefore, helpful in permitting the pumps or water treatment plants to work at a uniform rate, and they store water during the hours of no demand or less demand and supply water from their "storage" during the critical periods of maximum demand.

Reading Material Dams and Reservoirs for Multiple Purposes

Dams are important to the water industry because they control rivers, generate electric energy, and create major recreational venues. Their distinct services often benefit multiple parties, but their large impacts on natural systems draw fire from environmental activist groups. Although dams are developed for multiple water-management purposes and offer many benefits, they are inevitably controversial and can be the centerpieces of long-standing water disputes. In addition to political heat, these can generate expensive lawsuits and government decision processes that extend over many years.

For the most part the infrastructure of dams and reservoirs was developed by the utilities, local agencies, and industries that process and deliver water- related services in vertically integrated systems, but in many cases, dams were developed or assisted by federal agencies, mainly the U.S. Army Corps of Engineers (USACE), the U.S. Bureau of Reclamation (Burec) and the Natural Resources Conservation Service (NRCS). The companion feature to a dam is the lake or reservoir that it impounds. A reservoir is a lake where water is stored naturally or artificially and a pond is a small reservoir. In the United States, there are countless thousands of small lakes, compared to relatively few giant reservoirs. These small impoundments even extend down to the level of urban storm water detention ponds, which function the same way as flood control reservoirs, but hold less water and respond faster to inflows.

1. Evolution of dams and river infrastructure

Hydropower was based on stored water to turn water wheels for many centuries, and with the advent of electric power in the 1880s it was being developed to light cities in the United States. Then, the dam-building era began in earnest as engineers and business leaders saw the great possibilities of hydropower. By 1920, hydropower was providing 25 percent of the electric energy of a rapidly developing United States, and in the 1930s the creation of the Tennessee Valley Authority's system of dams was a centerpiece of FDR's New Deal. It harnessed a mighty river for economic and social development in an underdeveloped region. Along with development of dams came new lakes, land developments, and a water-based recreational industry that did not exist before.

Many dams were built in the United States between the 1930s and the 1960s, but few major dams or river structures have been built since the 1960s. This means that the youngest of America's major dams and navigation systems are passing the 50-year mark, and the same can be said for most developed countries, as well as many developing countries. This means that the legacy of aging dams will require vigilance and investments for renewal in the decades ahead.

Today, the United States has about 85000 dams that are large enough to warrant regulation, about 80000 megawatts of installed hydroelectric capacity, and systems of river management that extend even to small streams across the nation. The major issues are operation, maintenance, and renewal of these vital facilities. In the developing world, major dam sites remain attractive targets for new facilities.

2. Types and purposes of dams and reservoirs

The basic functions of dams, to control stream flows and impound water, create many situations, as stream flows occur from the smallest headwaters to giant bays and estuaries that empty into the ocean. In the United States, this involves big river and lake systems such as the Mississippi River and Great Lakes down to small headwater streams in hill or mountain areas.

Dams can be large or small, as measured by their height, width, and quantity of water impounded. The oldest design is the earthfill dam, which has an impervious core and upstream and downstream faces with pervious rock material. A rockfill dam is similar to an earthfill design, but it uses rock as a structural element and has an impervious membrane to seal water flows. A concrete gravity dam is normally a massive structure with a pervious foundation, a cutoff wall, a downstream and upstream apron, and anchor walls to aid in prevention of sliding. Slim, graceful arch dams are built of concrete but are much thinner than the concrete gravity dams. All dams have auxiliary features for required functions, such as outlet works for water flow and spillways to protect them. A dam normally has a service spillway for ongoing releases and an emergency spillway to release large storm flows. The size and capacity of these spillways can be contentious issues in dam risk analyses and lead to the need for large reinvestments for dam upgrades.

Most reservoirs serve multiple purposes, and in terms of numbers, the purposes of dams in the United States are (in order) recreation, farming, flood control, irrigation, water supply, mine waste retention, and hydro- power. The largest dams are usually those built with government funding and are for multiple purposes as outlined in the authorizing legislation.

The value added by a water-supply reservoir is to increase reliability of a surface water source of supply, which may even go dry during droughts. By capturing excess water during wet periods, the reservoir can equalize flows and release water for use when needed. As you can imagine, the saying "you will know the worth of water when the well goes dry" illustrates how valuable having a reliable source of water during dry weather can be.

Flood-control reservoirs work by capturing the high rates of flow during storm events and releasing the water slowly to avoid damage downstream. The Army Corps of Engineers has been responsible for many of these, such as the Cherry Creek Dam upstream of Denver, which protects the city from devastating flash floods.

A reservoir for irrigation is similar to one for municipal and industrial water supply in that it captures water during wet periods for release when needed. In contrast to municipal water supply, which meets human needs and can be practically priceless, the value of irrigation water supply is more variable. Irrigated farming usually depends on rainfall and supplementary supplies, and water may be required if the crop is to survive at all or it may be used to add to the yield of crops from doses of water to finish crops off at the end of a season.

Recreation is normally provided as a side benefit of other purposes of reservoirs, but it has tremendous value to people enjoying the water and to the local areas and businesses that benefit from it. Economists debate how to compute the benefits from recreation so as to assess its feasibility in justifying a water project, but normally the willingness to pay is insufficient to finance water projects only by recreation interests.

Navigation is facilitated by backing up water to increase the depths of river channels. In the United States, the main inland waterway system includes a number of dams, mainly constructed by the Corps of Engineers.

Lesson 2 Waterway

Inland waterways have always provided a link for agricultural products and manufactured goods from the interior to the coastal ports of the world. The evolution of inland waterways in the United States started with canoes and rafts then continued with river boats either rowed or pulled in natural rivers.

The type or types of waterways that could be developed will vary with local conditions. The types normally considered are open river, canalized streams with locks and dam, land-cut canals, or a combination of one or more of these types. Each type has its advantages and disadvantages that have to be considered.

1. Open river

The towing industry would prefer open-river navigation since it would eliminate delays normally encountered in passing through locks, but this is not practical on many streams because of their characteristics and local restraints. Many problems are associated with open-river navigation, and development and maintenance of this type of waterway usually involve some channel rectification, training and stabilization structures, maintenance dredging, and navigation aids. Changes in river stage and discharge produce changes in channel width and depth and in some cases channel alignment. The first cost of developing this type of waterway is generally less than that with other types but requires continuous surveillance and marking of the channel and considerable maintenance. Open-river navigation is maintained on the Mississippi River and Columbia River below Bonneville Dam.

2.Canalized streams

Canalized streams involve the construction of locks and dams to maintain adequate depths for navigation during periods of medium and low flows. Locks and dams would be required in streams having steep gradients with velocities too high for navigation or where conditions make it impractical to develop the required depths naturally because of rock outcrops, sediment movement, and other factors that could adversely affect navigation and flood-carrying capacity of the stream. Even with locks and dams, some channel improvement and regulating and stabilization structures and channel maintenance will be required. The principal disadvantage of this type is high initial cost and delays caused by tows passing through each lock. Canalized water was usually have lower velocities and greater channel width and depth through most of the reach of the pool during controlled river flows. Example of canalized waterways are the Ohio and Monongahela Rivers, the Mississippi River above St. Louis, MO, and Arkansas River. Locks might also be required in channels through estuaries, bays, near the mouths of some streams, and in some sea-level canals to prevent saltwater intrusion of minimize the effects of tides and differences in water levels with connecting waterways.

3.Canals

Land-cut canals have been used to connect two bodies of water, to bypass rock outcrops and rapids, and to reduce the length or curvature of the navigable channel. Canals can parallel existing streams or continue overland to reach specific destinations. Construction of canals can be expensive depending on the amount and type of excavation, land acquisition, and availability of disposal areas. When connected to an existing stream or other body of water, locks might be required in the canal. In order to reduce the amount of excavation, canals might be routed through shallow lakes and estuaries. Stabilization structures might be required along the banks of the canal to reduce erosion of the banks due to waves created by traffic and wind. Canals tend to be narrow and shallow to minimize cost and could be affected by surges resulting from lock filling or emptying when relatively high-lift locks are used. Examples of land-cut canals are the chain of rocks canal near St. Louis, MO, the New York Stage Barge Canal, the intracoastal canals, and the divide cut on the Tennessee-Tombigbee Waterway.

4.Basis of selection

Selection of the type of waterway adopted will depend on the amount and type of traffic that would be developed: characteristics of the equipment in general use; channel alignment and dimensions required; sedimentation problems to be resolved, safety, efficiency, and dependability; environmental effects; and comparative cost of construction, operation and maintenance.

5.Cost estimates

A series of layouts with cost estimates is needed to develop optimized costs. These life cycle cost estimates should include initial construction cost, maintenance cost, and replacement cost. Each of the layouts is required to move the required tonnage but each will have a different trip time. This trip time translated into benefits. The comparison of project costs versus benefits will provide the basis for section of the optimum layout. Generally, fewer higher-lift locks are cheaper than a greater number of lower-lift locks. Economy should consider both first cost and maintenance and operation cost without sacrifice of safety, efficiency, and dependability.

6.Basic project components

Navigation projects can be single purpose and only consider navigation, or a project may be developed for multipurpose such as flood control, hydropower, recreation, and water supply in addition to navigation. Therefore the basic components of a navigation dam could include:

1)spillway(gated or uncontrolled)

2)Overflow embankment or weir

3)Navigation pass

4)Lock or locks

5)Outlet works

7.Supplemental project components

The design of a single purpose or multipurpose project should accommodate each purpose as much as possible and develop a cost-effective functional plan. Common supplemental components are:

1)Powerhouse

2)Fish passage facilities

3)Recreation facilities

4)Water supply intakes

5)Water quality, low-flow controls, and multilevel outlets

6)Irrigation outlet work

8.Checklist for studies required

The development of waterways for navigation involves the study and evaluation of many factors to ensure efficiency, safe conditions, and reliability at minimum cost. Some of the studies and factors that should be considered in the planning and design phase are:

1)hydraulic and geological characteristics of the stream, and locations of existing bridges, highways, railroads, and industrial complexes;

2)Channel depths and widths available and requirements for anticipated traffic;

3)Need for channel realignment, training structures, and/or locks and dams;

4)Optimum locations for locks, dams, and port facilities if required;

5)Alignment and velocity of currents and movement of sediment in critical reaches and at proposed lock and dam sites;

6)Effects of various arrangements of lock or locks, dam and over flow weirs, and auxiliary lock walls;

7)Number and size of spillway gate bays and effects of overflow weirs and embankment on cost of project and non-navigation conditions;

8)Use of a navigable pass to reduce the height of lock walls;

9)Effects of structures on flooding ,overbank flow, and movement of sediment;

10)Effects of various types of lock filling and empting systems on navigation;

11)Effects of developments on water quality and local environment;

12)Feasibility of powerhouse installation and effects on navigation;

13)Feasibility of water conservation methods;

14)Effectiveness of various types of river training, stabilization structures, and navigation aids;

15)Navigation and flow conditions during construction;

16)Hydraulic model studies to determine:

(1)Optimum design for spillway and stilling basin operating under various conditions(full or partial width);

(2)Navigation conditions in lock approaches, best arrangement of locks, dam, and training structures, movement of ice and debris, and conditions during construction(comprehensive fixed bed or vessel simulator studies);

(3)Effects of structures on movement of sediment, channel development in lock approaches and in critical reaches, and scour with various cofferdam plans for construction(comprehensive movable bed);

(4)Conditions at other locations as needed such as harbor entrances, port facilities, assembly areas, and at bridges(fixed or moveable bed or vessel simulator studies);

17)Baseline environmental and water quality data collection and evaluation, and consideration of applicable environmental laws and regulations;

18)Relocations;

19)Historic and prehistoric preservation.

New Words and Phrase

1.interior *n.* 内部
2.tow *vi.* 拖；拉；牵引
3.optimization *n.* 最佳化，最优化
4.waterway *n.* 水运
5.open river 明渠
6.canalized stream 渠化河道
7.canal *n.* 运河
8.restraint *n.* 约束力；制约因素
9.maintenance *n.* 维护
10.rectification *n.* 纠正，修正
11.dredg *vi.* 疏浚
12.navigation aid *n.* 通航辅助设施
13.alignment *n.* 校直；调整
14.surveillance *n.* 监督，监视
15.outcrop *n.* 露出地面的岩层
16.parallel *adj.* 平行的；类似的
17.excavation *n.* 挖掘；开凿
18.acquisition *n.* 获得；购置物
19.disposal *n.* 闸门
20.erosion *n.* 侵蚀
21.surge *n.* 激增
22.sedimentation *n.* 淤积
23.layout *n.* 平面布置
24.replacement *n.* 代替
25.arrangement *n.* 安排；约定
26.assembly *n.* 装配
27.baseline *adj.* 基准的，基线的
28.preservation *n.* 保存，保护
29.powerhouse *n.* 发电厂房
30.feasibility *n.* 可行性

Notes

1.Many problems are associated with open-river navigation, and development and maintenance of this type of waterway usually involve some channel rectification, training and stabilization structures, maintenance dredging, and navigation aids.

明渠通航存在很多问题，该类航道的发展和维护主要涉及航道，整治和稳定建筑物，维护性疏浚以及通航辅助设施。

associate with 与……交往，联系；交接

A range of earthwork options is commonly used to manage or alleviate problems associated with excess water in a catchment or on the farm.

involve 涉及，牵扯等，后面通常接结果

2.Another procedure being studied involves the application of the shell theory to arch dam analysis.

当河道的比降和流速较大，或者由于礁石、泥沙运动以及其他对通航条件和行洪能力产生负面影响的因素无法达到要求水深的时候，则需要修建船闸和大坝。

adversely *adv.* 不利地；逆地 rock outcrops 指礁石

Salts are left in the root zone as a result of evapotranspiration and can adversely affect the soil structure and limit plant growth.

3.Navigation projects can be single purpose and only consider navigation, or a project may be developed for multipurpose such as flood control, hydropower, recreation , and water supply in addition to navigation.

航道工程可以仅为通航服务，或者兼顾其他综合目标，比如防洪、发电、休闲以及供水。

navigation project 航道工程 in addition to 除……之外

Where feasible, the provision of an emergency spillway in addition to permanent service spillways may result in reduced cost of construction as well as an increased factor of safety against overtopping of the dam.

Comprehensive Exercises

I.Answer the following questions on the text.

1.How many type of waterways, what are they?

2.How to solve the navigation problem in the open channel?

3.When do the locks and dams should be constructed in the canalized streams?

4.What are the controls to determine the type of waterways?

5.What are the components in the navigation projects?

II.Fill the most appropriate words or phrases in the correct forms in the blanks from the list below.

vary	depend on	involve	channel dimension	erosion
Supplement	regulate	minimize	conservation	in general

1.A water supply system, for example, _____dams and other structures as well as the flow and storage of water.

2.For example, a given volume of soil _____behavior depending on the predominant size of the particles.

3.The size of the foundation_____the allowable compressive stress under the foundation and on considerations related to the casting technique.

4.______required for navigation will depend on channel alignment, size of tow, and whether one-way or two-way traffic is to be accommodated.

5.The exact design of the_____protection against the main ship propellers and the bow and stern thrusters action, is very difficult.

6.The level of water within the lock _____so that shipping can be raised or lowered to different elevations.

7.In urban areas, one roadway is sometimes placed above the other in a double-deck configuration to ______ land needs.

8.For the time being, water ______practices have been taken in an attempt to control groundwater abstraction.

9.The two types of reaction turbines ______ use are the Francis turbine and the propeller turbine.

10.These background levels of impurities______by human activities.

III.Translate the following sentences into Chinese from the text.

1.These limitations on navigation prompted the development of controlled waterways with locks and dams. The dams created reservoirs with predictable channel depths and the locks provided a means to allow safe passage from one pool to the next.

2.Open river navigation implies the use of natural streams for navigation without locks and dams the development of open-river navigation usually involves lower first cost but maintenance cost could be high because of the complex nature of these streams, their tendency to meander and migrate, and the difficulty of designing the training and stabilization structures needed.

3.Locks and dams would be required in streams having steep gradients with velocities too high for navigation or where conditions make it impractical to develop the required depths naturally because of rock outcrops, sediment movement, and other factors that could adversely affect navigation and flood-carrying capacity of the stream.

4.Construction of canals can be expensive depending on the amount and type of excavation, land acquisition, and availability of disposal areas. When connected to an existing stream or other body of water, locks might be required in the canal.

Reading Material River Training Works

The improvement of natural rivers for navigation involves channel realignment, stabilization, training structures, and in many cases the modification or replacement of existing bridges. In streams carrying large quantities of sediment, a sinuous channel is easier to develop and maintain than channels in long straight reaches or long flat bends and should be considered in the layout and planning for the project. The sinuosity of a stream varies over a wide range. However, design should be based insofar as practical on the alignment of reaches that have been reasonably stable with a channel adequate for the traffic anticipated. Channel alignment involves corrective dredging, training and stabilization structures, or in some cases cutoffs.

1.Dredging

Corrective dredging is used to realign the channel or bank lines and to develop cutoffs.

Dredging in the channel bed involves in the removal of erosion-resistant material such as gravel bars, rock outcrops, or clay plugs. Usually dredging within the channel bed without some training or contracting structures will produce only temporary results and might have to be repeated after each high-water period or significant rise in river stages.

2.Channel stabilization

Channel stabilization involves the protection of the banks of streams or canals from erosion caused by currents or wash from waves created by wind and traffic. Natural streams with erodible bed and banks will tend to meander and migrate and, unless this tendency is resisted, will be constantly changing .erosion of the channel bed along a bank will tend to undermine the bank or steepen its slope to the point that caving or sloughing of the bank occurs. Erosion and caving of banks can adversely affect channel alignment and depth, can increase sediment load and maintenance cost, and could result in the loss of valuable land and endanger local installations such as buildings, rail lines, highways, bridges, docking facilities, and flood-control levees or floodwalls.

Bank protection can be a major cost in the development of a waterway for navigation and should be considered during the initial planning of the project. Some of the cost might be considered as part of the flood-control aspects, particularly if it is a multipurpose project. The type of bank protection vary depending on the characteristics of the stream, particularly the variations in stage and discharge and the erodibility of the stream bed and stream banks. Bank protection and stabilization might consist of structures such as dikes designed to divert currents away from the bank or improve the alignment and velocity of currents along the bank. The most common type of bank protection is some type of revetment covering the bank and channel along the toe of the slope with erosion-resistant material or blanket.

3.Training structures

The development and improvement of channel alignment and depth often require the use of structures to reduce the width of the channel, realign the channel, and stabilize the low-water channel. These structures usually consist of some type of dikes constructed of timber pile clusters, stone, or piling with stone fill. The type of structures used and their arrangements should be based on the characteristics of the stream, problem design of the structures should consider the effects of the structures on currents existing in the reach and the movement of sediment, and the effects of the resulting currents on navigation.

1) Spur dikes

The most common type of structure used in channel improvement and development extends from the riverbank channel ward in a direction approximately normal to the channel being developed. These dikes are usually included in a system of two or more and are generally referred to as spur dikes. These dikes have also been referred to as transverse dikes, cross dikes, wing dams, jetties and so on. Spur dikes should be designed to provide a favorable lateral differential in water level to increase their effectiveness and reduce construction and maintenance cost. Improperly designed spur dikes could be either ineffective or unstable, increase channel losses, cause shoaling upstream, or develop a channel of poor alignment.

2) longitudinal dikes

Longitudinal dikes are continuous structures extending from the bank toward the downstream generally parallel to the alignment of the channel being developed. Properly designed longitudinal dikes are the most effective type of structure in developing a stable channel but are also the most expensive. These structures can be used to reduce the curvature of sharp bends and to provide transitions with little resistance or disturbance to flow. However, once in place, it is difficult and expensive to change the alignment of the dike. If not constructed high enough, flow over the top of the dike would tend to move tows toward and against the structure. Little ir no natural deposition can be expected landward of the dikes but the area could be used for the placement of dredged material. Consideration for fish and wildlife might indicate the need for modification, such as providing openings to develop and maintain water areas behind the dike.

3) Vane dikes

Dikes placed in the form of a series of vanes have proved effective as a means of controlling channel development and sediment movement under certain conditions. These dikes consist of lengths of dikes located out from the bank with space in between and placed at a slight angle to the alignment of the currents to develop the lateral differential in water level desired. The length of the gaps between the dikes is usually about 50%-60% of the length of each vane and should be placed where there is or will be movement of sediment. These dikes, developed as a result of model studies, have been used successfully on the Mississippi River. Vane dikes can be used independently or as extensions to spur dike system. Vane dikes are much cheaper than conventional dikes since they can be placed in relatively shallow water generally parallel to the channel limit line and produce little disturbance to flow.

4) Closure dikes

In reaches where there are islands and divided flow, depths will tend to be limited. In such cases, it will be necessary to reduce or eliminate the low and medium flows from all the principal channel being developed. This is accomplished by diverting sediment into the side channels or by closure structures across the side channels. Sediment can be diverted into the side channel by developing the lateral differential in water level with spur dikes, vane dikes, or by a combination of both. When the length of the side channel is short relative to that of the main channel as in a bend, closure dikes across the shorter channel will tend to be difficult to maintain because of the high head that develops across the dike, producing scour downstream. In such cases, one or more closure dikes placed downstream at successively lower elevations will tend to divide the total drop between the dikes and reduce the amount of scour that would tend to endanger a single structure.

Lesson 3 Port Engineering

The function of a port is to provide an interface between two modes of transport-land and sea-for cargo and passengers. The requirements for sea transport are: (1) an adequate area of water of sufficient depth for navigation and berthing; and (2) adequate shelter so that berthing, loading and unloading can be carried out safely and efficiently. The requirements for the land use are: (1) adequate land area for working space, loading and unloading vessels and for handling and storage of cargoes; and (2) suitable access to areas served by the port

1. Siting of ports and harbors

The siting of a port is generally dictated by commercial and economic requirements, particularly in relation to land transportation. A natural harbour is to be preferred in order to avoid the necessity of expensive breakwaters, even though some dredging may be required to provide the necessary area of deep water. If the material to be dredged is suitable, land reclamation may be possible using the dredged material to provide land for the shore facilities of a port.

If a natural harbour is not available, breakwaters will be required to provide adequate shelter. Breakwaters are normally very expensive however, and this must be weighed against any additional transport costs and compared with the expenditure incurred at a port where breakwaters are not required.

In planning a new harbour involving breakwaters, consideration must be given to the following factors, in addition to the design of the breakwater itself: (1) waves; (2) littoral drift and sedimentation ;(3) tides and currents; and (4) navigation.

1) Design of harbors

The main purpose of breakwaters is to provide protection from waves, and the biggest wave reduction is effected with the smallest entrance sited remote from the direction of approach of the waves. However, this can cause difficulty when approaching the entrance with heavy seas abeam the vessel. As harbours are normally designed to serve as a harbour of refuge, i.e. a protection to be sought by vessels during the height of a storm, it is common to site an entrance at a small angle to the heaviest sea, thereby improving accessibility at the expense of smoothness within the harbour.

Wave-height reduction within a harbour is improved as the distance from the entrance, and the width parallel to the shore, increase. It is desirable to have wave-spending beaches or armoured slopes which absorb wave energy facing the waves within the harbour, rather than vertical walls which reflect waves and could cause resonance resulting in significant increases of wave heights. Wave heights within a harbour are normally predicted using numerical models or a physical model; in both cases, various breakwater alignments can be tested to give the optimum alignment. An empirical method for assessing wave heights within a harbour is given in the Stevenson formula:

$$h_{\mathrm{p}} = H\left[\left(\frac{b}{B}\right)^{\frac{1}{2}} - 0.027D^{\frac{1}{4}}\left(1+\frac{b^{\frac{1}{2}}}{B}\right)\right] \tag{2.1}$$

where h_{p} is the height of reduced wave at any point in the harbour, H is the height of wave at entrance, b is the breadth of entrance, B is the breadth of harbour at P, being length of arc with center at midway of entrance and radius D and D is the distance from entrance to point P.

2) Sedimentation

Sedimentation in a harbour can arise from three sources:(1) littoral drift; (2) tidal movements; and (3) where a harbour is located at a river mouth, from the river. The minimizing of sedimentation in navigation channels, at the entrance and within the harbour, is of prime importance in reducing the cost of maintenance dredging.

Littoral drift occurs to some extent along most coastlines. If the path of the drift is obstructed by a solid structure, the heavier particles will accumulate on the drift side and this accumulation may well extend round to the inside. The finer particles of the drift, which outside the harbour are kept in suspension by current velocities will, on entering the harbour, no longer be maintained in suspension and will settle out. Littoral drift normally occurs in one direction, but at certain times of the year or under some storm conditions, the direction of drift can be reversed.

Where a harbour is subjected to large tidal ranges, material in suspension will be brought into the harbour as the tide rises and, during periods of slack tide, material will settle on the sea bed. Where a harbour is at a river mouth, the material carried down by the river is a further source of sedimentation. The interaction of river flows and movements of the sea makes for further complications, with the added difficulty of the difference in density between fresh and salt water.

Predictions of sedimentation are best carried out by numerical modelling. Physical models can also be used, but these can be less accurate particularly with fine material in suspension because of the difficulty of scaling-down the fine particle sizes to the scale of the model, and results should be treated with caution.

2. Port planning

The planning of a new port or expansion or improvement of an existing one requires many factors to be taken into consideration. Apart from passenger ferry terminals and cruise ship terminals, ports are primarily provided for the handling of cargo. Among the factors to be considered are:

(1)Nature of cargoes to be handled.

(2)Sizes and types of vessels to be catered for.

(3)Method of cargo handling.

(4)Land area and operations.

(5)Land access.

1) Types of cargoes

Between 1960 and 1980 a major revolution in the handling and carrying of maritime cargoes took place and this has led to new concepts in the design of ships, ports and land transportation systems. Generally speaking, during this period emphasis was given to handling and carrying cargoes in larger units, e.g. containers in the case of general cargo, and larger single shipments of bulk commodities such as wheat and oil, etc. Ship sizes also increased to obtain the benefits of the increased scale of operation

Nonunitized (or break bulk) cargoes. These consist of small consignments requiring to be handled individually. The volumes now being conveyed by this method are rapidly diminishing and nonunitized working is practiced only in areas where labour is plentiful.

Unitized cargoes. Unitization of cargoes permitting larger units of general cargo to be handled by mechanical equipment, so replacing labour, has become attractive. These are sealed units, capable of being lifted from the bottom by fork-lift trucks or from the top at the ISO four corner lock attachments by cranes and mobile equipment. They are also stackable. Specialized ISO containers have been developed as refrigerated and liquid tank units, but all are to the standardized overall dimensions and equipped with the ISO universal handling devices. Some of these, e.g. refrigerated units, require support services in the way of electrical power whilst in transit through the port.

2) Bulk cargoes

Bulk cargoes fall into two categories: (1) dry; and (2) liquid. Commodities of these types, more often than not, are shipped in purpose-built vessels or carriers and are loaded and unloaded using specialized berths or terminals equipped with mechanical handling systems suitable for the commodity being handled. Typical commodities are grain, mineral ores, timber, sugar, vegetable oils, mineral oil and petroleum products, liquid, chemicals, liquefied petroleum gases (LPG) and liquefied natural gas.

3.Sizes and types of vessels to be catered for

Ships are classified under a number of tonnages as follows.

(1)Gross registered tonnage (GRT): The value derived from dividing the total interior capacity of the vessel by 2.83 m^3, subject to the provisions of applicable laws and regulations.

(2)Net registered tonnage (NRT): The gross tonnage of the vessel minus the tonnage equivalent of crew cabins, engine-rooms, etc.

(3)Displacement tonnage: Indicates the total mass of the vessel, and is obtained by multiplying the volume of the displaced sea water by the density of sea water (1.03 t/m^3).

(4)Dead weight tonnage (DWT): Dead weight of a vessel is the weight equivalent of the displacement tonnage minus the ballasted weight of the vessel. Consequently, it indicates the weight of the cargo, fuel, water and all other items which can be loaded aboard the vessel.

(5)Tonne measurement: The value derived from dividing the cargo spaces of a vessel by 1.13 m^3.

For port engineering purposes, DWT is the most significant although, for calculating berthing energies, the displacement of the vessel is required. The shipping industry uses the long ton. This is almost the same as the metric tonne and for planning purposes can be treated as being interchangeable.

4.Types of vessels

Vessels are generally categorized by the types of cargo they handle as follows.

(1)*General cargo.* These generally carry nonunitized (breakbulk) cargoes and/or unitized cargoes, but can also carry some containers. These range in size from small coasters (2000-3000 DWT) to long-distance vessels up to 30000 DWT.

(2)*Container vessels.* These are specially designed ships for the purpose of carrying containers and can vary from small feeder vessels carrying perhaps 150 TEU up to the very large container vessels (used on long sea routes) carrying up to 4000 TEU and being of about 70000 DWT.

(3)*Roll-on roll-off vessels.* These are specially designed to allow the movement of cargo through stern or bow ramps by vehicular movements without the need for cranes or other lifting devices, and are generally used on the shorter sea routes.

(4)*Bulk-cargo vessels.* These are normally designed specifically for a particular trade, such as iron ore, coal, grain sugar, etc. and can range from small vessels of 20000 DWT up to large bulk carriers of up to 60000 DWT.

(5)*Tankers.* These are designed for liquid bulk cargoes and can range from small vessels of 20000 DWT up to the very large oil tanker of up to 1 million DWT.

Certain characteristics of vessels may also need to be taken into account. Some vessels are equipped with bow thrusters for ease of manoeuvring, and these have been known to cause damage to quay walls. Problems can also occur with vessels that have bulbous bows, where the projecting bow located below water can cause damage to piled structures.

In planning a port development, knowledge of the following characteristics of vessels likely to use the port is required in addition to the dimensions of vessels (length, beam and draft).

(1)Ship layouts, including the locations and dimensions of ramps and hatches, loaded and unloaded deck heights, superstructure positions and clearances for dockside cranes.

(2)Handling characteristics of ships for manoeuvring and turning operations.

(3)Windage areas of ships to assess forces on berths.

(4)Ship mooring line sizes and capacities for bollard pulls.

(5)Deck crane capacities and reaches.

5.Land area

This depends on: (1) throughput of cargo; (2) type of cargo; (3) methods of cargo handling; and (4) length of time cargo remains in the port. A modern general cargo berth is normally 200m long and 200m or more deep. Thus, an area of 200 m × 200 m, or 4 ha, is required. With efficient cargo handling, this will handle approximately 250000 t of cargo per year. A container berth requires more land behind the berth to maximize the throughput. Container berths are generally 300 m long or greater and with up to 200 to 800 m depth, although this can be reduced if containers are stacked. The area can therefore range up to about 20 ha which would handle up to about 1 million t of cargo per year. However, the land requirements must be investigated for individual cases according to the factors mentioned above. With a general cargo area, part of the land will be utilized by transit sheds and warehousing. In a container berth, the land area will largely be open for storage of containers with sheds for filling and emptying containers, unless these operations are carried out at an inland depot away from the port.

6.Other considerations

Other factors requiring consideration in the planning of port developments include:

(1) Tugs and pilotage.

(2) Security and policing services.

(3) Fuel bunkering facilities.

(4) Equipment maintenance facilities.

(5) Services to ships-water, electricity, sewerage, telephone.

(6) Rest rooms, canteens and offices, etc.

(7) Post offices.

(8) Customs and immigration arrangements.

New Words and Phrase

1.cargo *n.* 货物

2.berth *n.* 泊位

3.dictate *vi.* 口诉，规定

4.breakwater *n.* 防波堤

5.shelter *n.* 居所； 住处

6.littoral *adj.* 沿岸的

7.approach *n.* 方法，途径

8.refuge *n.* 庇护，避难

9.accessibility *n.* 可（易）接近性；可访问性

10.breadth *n.* 宽度

11.reverse *adj.* 相反的

12.emphasis *vi.* 强调，重视

13.nonunitized *adj.* 非集装箱的

14.unitized cargoes　集装箱货物
15.whilst　*adv.*　在……期间
16.bulk cargoes　*n.*　散装货物
17.interchangeable　*adj.*　可交换的
18.roll-on roll-off vessel　滚装船
19.thruster　*n.*　推进器
20.manoeuvring　*n.*　手段
21.ramp　*n.*　坡道
22.hatch　*n.*　舱口
23.bunker　*n.*　地堡；掩体；煤舱

Notes

1.The requirements for the land use are: (1) adequate land area for working space, loading and unloading vessels and for handling and storage of cargoes; and (2) suitable access to areas served by the port.

港口修建的陆地要求主要为：(1) 可用于作业的足够的空间，轮船的装货或者卸货，以及货物的处理和存储；(2) 合适可进入港口服务区域的通道。

access to　接近，去……的通路，使用……的机会

The access to machines is through the hatch removable covers being provided over the machines.

2.In planning a new harbour involving breakwaters, consideration must be given to the following factors, in addition to the design of the breakwater itself: (1) waves; (2) littoral drift and sedimentation; (3) tides and currents; and (4) navigation.

在规划涉及防波堤的港口阶段，除了涉及防波堤自身需要考虑的因素外，还需要考虑下述因素：(1) 波浪；(2) 离岸流和泥沙淤积；(3) 潮汐和洋流；(4) 通航条件。

in addition to　除了　　littoral drift　沿岸流

In addition to the benefits, however, damming the Nile caused a number of environmental issues.

3.The planning of a new port or expansion or improvement of an existing one requires many factors to be taken into consideration. Apart from passenger ferry terminals and cruise ship terminals, ports are primarily provided for the handling of cargo.

规划新港口或者扩充或改善已有的港口阶段，需要考虑很多因素，除了作为乘客渡船的终点站和大型游轮的终点站外，码头的首要任务是用于处理货物。

take into consideration　考虑　　apart from　除了

Apart from the water availability in the topsoil, the evaporation from a cropped soil is mainly determined by the fraction of the solar radiation reaching the soil surface.

Comprehensive Exercises

I.Answer the following questions on the text.

1.What should we consider in planning a new harbour involving breakwater?

2.How do we predict the wave height?

3.How do we estimate the sedimentation?

4.What are the types of the cargo?

5.What do the land areas depend on in the port planning stage? What are the types of vessels to be catered for?

II.Fill the most appropriate words or phrases in the correct forms in the blanks from the list below.

in relation to	approach	reverse	be subjected to	bring into
take into	fall into	according to	sedimentation	throughput

1.______the data of different authors an annual runoff from Antarctica and Greenland is estimated.

2.Heights are given______other heights, for example, the usual procedure is to give the elevation above sea level.

3.The rate at which soils______ an average field capacity is slower with clayey soils than with sands.

4.The precipitated chemicals, original turbidity, and suspended material are removed from the______basins and from the filters.

5.The Port reaches an annual______ of 6.5 million tons.

6.Government officials have said the coefficient was not accurate in reflecting China's inequality level, as it does not ______of the country's regional gap in consumption.

7.Salinity spread in the catchment will be stopped and possibly_______the trends of rising water levels in the upper parts of the catchment.

8.The loads a structure_______are divided into dead loads ,which include the weights of ail parts often structure and live loads with are to the weigh movable equipment etc.

9.During this period, the power station ton the left bank and the permanent ship lock______.

10.Bulk cargoes______two categories: (1) dry; and (2) liquid. Commodities of these types.

III.Translate the following sentences into Chinese from the text.

1.The function of a port is to provide an interface between two modes of transport-land and sea-for cargo and passengers. The requirements for sea transport are: (1) an adequate area of water of sufficient depth for navigation and berthing; and (2) adequate shelter so that berthing, loading and unloading can be carried out safely and efficiently.

2. Where a harbour is subjected to large tidal ranges, material in suspension will be brought into the harbour as the tide rises and, during periods of slack tide, material will settle on the sea-bed.

3.These will depend largely on the nature of the cargoes and the types of vessels likely to use the port. The most important consideration is whether dockside cranes are required or whether ships' own lifting gear will be used for loading and unloading. Apart from cranes, cargo-handling equipment can range from fork-lift trucks, which can have a capacity from 30 to 400kN for general cargo, to special container-handling equipment.

4.However, the land requirements must be investigated for individual cases according to the factors mentioned above. With a general cargo area, part of the land will be utilized by transit sheds and warehousing.

Reading Material General Select from Port Design

The open berth structures constitute with their quay platforms a prolongation over the slop from the top of the filled area out to the berth front. In this chapter only berth structures in reinforced concrete or platforms in reinforced concrete founded on concrete — filled tubular steel piles will be described. Open berth structures built of wooden materials will not be described but construction principles, load bearing capacity etc. are largely the same as for structures in reinforced concrete. In the same way as for solid berth structures the open berth structure can also be divided into two main types, depending on the principles according to which the front wall and the platform are designed to resist the loading so that the berth structures get the necessary stability.

-Lamella quays:

The quay platform and the front wall are founded on vertical lamellas, which provide the loaded berth structure with a satisfactory stability. The berth structures are stable enough in themselves to resist loads from ships, useful loads, possible pressure from fill at the rear of the structure, etc. without anchoring of the structure. In the same way as for gravity wall structures, one can build the lamella quay itself first and then fill behind close up to the structure.

-Column or pile quay:

The quay platform and the front wall are founded on columns or pile which do not have a satisfactory stability against external forces. Therefore the quay structure must be anchored, for instance by a friction plate in the filling. The structure must anchor the quay structure, for instance. The structure must then be built simultaneously with filling or preferably after the fill has been established.

Very often the two types are combined. For instance, in a pier or jetty type of berth structure the shore base is anchored by a retaining wall while at the head of the jetty the horizontal loading is resisted by lamellas. The quay platform between the shore base and the head of the structure is founded on columns and /or piles.

Generally, thc cconomically best solution for an open berth structure will be obtained when the total width B is as small as possible in relation to the height of the berth front. The different characteristic dimensions influencing the structural dimensioning and design are discussed below.

H-The height of the berth front is determined by the necessary water depth and height of the berth surface above LAT which is the usual reference level for berth structures.

H_1-The depth between LAT and the bottom of the harbor basin is determined by the draft of calling ships when fully loaded plus an over depth to cover the trim of the ship, wave height and safety margin against sea bottom irregularities.

H_2-The elevation of the top of quay platform above LAT is determined by the contour lines of the area behind the quay or by types of ship, which will call at the berth structure. Top of platform should not be at a lower level than the highest observed water level plus 0.5 m.

H_3-This distance is determined by the location of the rear wall or the retaining slab above LAT. The bottom of the slab should not lie lower than mean sea level i.e. Z_0 above LAT. The rear structure should be placed at such a level that concreting can be carried out on dry land. In case it is necessary for the rear structure to lie lower, the construction can be carried out with pre-fabricated elements placed on a leveled base under water. By increasing the height *H* and the angle of the slope, the total width *B* will be reduced, but the economic benefit by reducing *B* can easily be eliminated by higher installation costs because of underwater work.

a-The slope should start about 1.0 m behind the berth front so that the toe of the slope is kept away from the turbulence caused by propellers. Possible rocks falling off the slope will then probably be prevented from falling outside the berth front line.

b-The distance b is determined by the steepness of the slope. The angle α will normally vary between 38.7° (1:1.25) and 29.7° (1:1.75), depending on the materials of which the slope is made, whether covered with blasted rock etc. Usually the angle is 33.7° (1:1.5).The angle of the slope is determined by the stability of the slope itself, the coarseness of the materials and the danger of erosion from waves and propeller turbulence.

c-This is a very exposed point because of the danger of slides in front of the rear wall or retaining friction slab. The width C of the shoulder must be at least 1.0-1.5 m, and the shoulder itself must be well covered. The width C must also be sufficient to ensure the stability in front of the retaining slab.

d-The width of the anchor or friction slab which shall resist the horizontal load depends on the frictional angle in the ground, the safety factor wanted against sliding and the vertical load acting on the retaining slab.

e-The distance between the berth line and the center line of the first row of supporting columns or piles is determined by the possibility that a ship with long bulbous bow protruding well forward of the forepeak below the water line and with markedly flared hull at the bow like larger container ships, should hit the columns or piles if berthing with a large berthing angle. To avoid this possibility the centerline of the columns or piles should lie about 2 m behind the berth line.

Lesson 4 Irrigation and Drainage

1. Irrigation

Irrigation is desirable where natural rainfall does not meet the plant water requirements for all or part of the year. Irrigation is essential for agriculture in the desert but even in areas such as northern Europe it can improve the yield of crops normally grown under rainfall conditions only.

1.1 Soil moisture

The soil can be considered a moisture reservoir. Soils can be classified under the International Soil Science Association (ISSA) system. Water is held by the soil in the soil pores. The amount of water held can be defined as follows:

(1)Saturation: the state of complete soil wetness when no further water may be added to the soil.

(2)Field capacity (FC): the condition reached after water has drained from the soil by gravity.

(3)Permanent wilting point (PWP): the condition reached after plants have extracted all the moisture they can from the soil.

(4)Available water: defined as (FC - PWP), the amount of water held by the soil that plants can use.

Plants respond to how tightly the water is held by the soil which is defined as soil moisture tension. Generally, it is assumed that the soil moisture tension at field capacity is 0.3 bar pressure. Soil moisture tension at PWP is assumed to be IS bar.

With knowledge of the crop rooting depth, the available soil moisture and the crop water requirements, it is possible to select a suitable irrigation interval (time between irrigations). Not all water in the root zone is readily available to the crop. It is normal to allow the crop to deplete only 50% of the available moisture before irrigating again.

1.2 Crop water requirements

Crop water requirements are defined as the depth of water required to meet the water loss through evapotranspiration (ET_{crop}) of a crop. The effect of climate on crop water requirements is given by the reference crop evapotranspiration (ET_0) which is defined as the rate of evapotranspiration from an extensive surface of green grass of uniform height (8 to 15 cm):

$$ET_{crop} = k_c \cdot ET_0 \tag{2.2}$$

where k_c is the crop coefficient which varies with crop, growth stage, growing period and prevailing weather conditions.

The most reliable method of estimating ET_0 is generally considered to be the PENMAN method. This method is best described by Doorenbos and Pruitt which also gives details of crop coefficients for a wide range of crops .Values of ET_{crop} are normally calculated for 10-day periods. A simpler method was proposed by Blaney and Criddle in which the monthly crop water requirements ET_{crop} (in millimetres) are found by multiplying the mean monthly temperature T_m (℃) by the monthly percentage of annual daytime hours

A simple calculation of gross irrigation requirements (I_{gross}) can be made as follows:

$$I_{gross} = \frac{ET_{crop} - R_c}{E_a} \tag{2.3}$$

where R_c is the effective rainfall and E_a is the field application efficiency.

1.3 Irrigation efficiency

It is necessary to account for losses of water incurred during conveyance and application to the field. Efficiencies can be divided into three parts: (1) field application; (2) field canal; (3) distribution efficiency. E_a is dependent on soil type and type of irrigation system used. E_f is dependent on type of field channel used and area served. Distribution efficiency (E_d) is dependent on area served, the level of water management, and canal seepage which is the main component.

1.4 Irrigation methods

The choice of method of irrigation is dependent on technical feasibility and economics. Normal methods fall into four main categories: (1) surface; (2) trickle; (3) subirrigation.

1.4.1 Surface irrigation

Surface irrigation is still the most common method of irrigation employed and is suitable for the irrigation of most soils with an infiltration rate of less than 150 mm/h and for lands with a flat topography with an overall slope of less than 3%, although these limitations are exceeded in some situations. There are four main types of surface irrigation: (1) basin; (2) border strip; (3) furrow; (4) corrugation irrigation.

1) Basin irrigation

Water is applied from a small canal by gravity to fill a level basin surrounded by earth bunds. In practice, these basins are often small but the most efficient irrigation is obtained by using large basins, at least a hectare in area and accurately levelled. Water should be applied to those basins at a rate of at least 2 to 4 times thc infiltration rate of the soil.

Basin irrigation is most suitable for very flat and level land and soils with low infiltration rates. When adopted for uneven topography the basin size must be kept small in order to limit the quantity of land levelling required. Land levelling rates in excess of 1000m^3/ha should be avoided.

2) Border strip irrigation

The land is divided into strips separated by earth bunds which run generally down the slope, and water is applied at the head of the strip and allowed to flow down the slope infiltrating the soil as it flows across it. The strip is graded at an even slope along its length in the direction of flow and level across the strip. Efficient irrigation is obtained by choosing the strip width, length and discharge to meet the soil infiltration rate and land slope conditions to give as constant a depth of water as possible infiltrated over the length of the strip.

3) Furrow irrigation

Furrow irrigation is used for the irrigation of row crops or crops grown on beds between furrows. Furrow irrigation usually implies sloping land although horizontal furrows can be used for row crops within level basins. Water is applied to the upper end of each furrow and flows down the furrow with water infiltrating into the beds between the furrows on which the crop is grown. Furrow spacings are a function of crop and type of tillage machinery used. Typically furrows are spaced 0.75 to 1.05 m apart.

4) Corrugation irrigation

Corrugation irrigation is a variant of furrow irrigation in which the furrows are very small. It is suitable for medium soils only and is used for close growing crops such as wheat. The Corrugations are some 10 cm deep and spaced 40 to 75 cm apart. Because the corrugation flows are small, slopes up to 8% have been used. This method of irrigation is not widely used outside the US.

1.4.2 Trickle irrigation

The basis of trickle irrigation is to provide irrigation water to individual plants. A plastic pipe is run along the ground at the base of a row of plants and water is carried to each plant through orifices in the pipe or using an emitter. Trickle irrigation is more accurately described as localized irrigation as it includes a wide range of emitters such as micro-sprinklers and bubblers. Trickle irrigation is most suitable for row crops and trees and is generally able to use more saline water supplies than surface irrigation or sprinkler irrigation.

1.4.3 Sub-irrigation

Sub-irrigation is only suitable for specialized soil conditions. High horizontal permeability and low vertical permeability are required, or a barrier layer beneath the root zone. Water is passed to the crop from open feeder ditches via buried perforated pipes. Control of the water level in the ditches determines the quantity of water available to the crops. A combined system of irrigation and drainage is common with the ditches and pipes doubling for both irrigation and drainage.

2. Drainage of agricultural land

Agricultural drainage is necessary to remove excess water from the soil to improve the agricultural potential. The benefits of drainage may include:

(1)Seed germination—excess moisture associated with low temperatures impairs germination. Waterlogging may cause seeds to rot and not germinate.

(2)Crop growth—most crops require air in the root zone to grow.

(3)Control of water table—high water tables will limit depth of root zone.

(4)Disease — waterlogged crops are more susceptible to disease.

(5)Yield gain — generally higher crop yields are experienced from drained land.

(6)Poaching — wet soil that carries stock experiences surface damage by grazing animals.

(7)Cultivation — improved drainage will allow easier access for cultivation machinery.

(8)Salinity — control of salinity in crop root zone.

Drainage systems can be defined as subsurface and surface. Surface drains are designed to remove excess runoff from the land which would otherwise cause localized flooding. Subsurface drainage is designed to remove excess water from the soil mass.

2.1 Sub-surface drainage of irrigated land

Sub-surface drainage for irrigated lands in arid areas is normally associated with the control of the water table depth. Most crops grow best with the water table below their root depth although crops may not be affected by a higher water table for a short period. Rice is an exception since it grows well in totally waterlogged conditions.

The necessary drainage is frequently achieved by providing perforated drainage is frequently achieved by providing perforated drainage pipes below ground at regular intervals.It is necessary to install the drains below the desired design water table depth.The shallowest drain for water table control is:

$$H+0.5h+0.1\text{m} \tag{2.4}$$

where H is design water table depth given above and h is rise in water table resulting from the maximum individual recharge from a water application.

2.2 Drainage surplus

The quantity of water to be removed by a subsurface drainage system can be estimated from a water balance:

$$Q_s = R_f + S_e + S_i - D_n \tag{2.5}$$

where Q_s is water to be removed by drainage, R_f is recharge to the water table from rainfall or irrigation, S_c is seepage from canals or rivers, S_i is groundwater flow into the area and D_n is groundwater flow out of the area. Recharge (R_f) to the water table will vary with soil type, irrigation method and efficiency of water management.

Groundwater inflow and outflow can be calculated from data on groundwater slope, flow cross-section and soil permeability using Darcy's law, which states that:

$$V = \frac{Kh}{L} \tag{2.6}$$

where V is flow velocity in metres per day, K is hydraulic conductivity of the soil in metres per day, and h/L is hydraulic gradient.

2.3 Drainage of lands subjects to excess rainfall

The drainage of irrigated land in arid areas is described above. However, many areas require drainage due to an excess of rainfall. The drain discharge due to rainfall rises to a peak following a rainstorm and then recedes.

For the design of a buried pipe-drainage system, the discharge is often based directly on rainfall data. For instance, in the UK field drainage design is based on 5-day rainfall divided by 5 to give the daily drainage rate with return periods. Typical drainage rates in northwest Europe would be of the order of 7 to 10 mm/day. Drainage systems incorporating mole drainage are normally based on a 1-day rainfall value, because of the shorter response.The design depth to water table is often taken at 0.5 m for shallow rooted crops and 0.75 to 1 m for deep-rooted and high-value crops. Drains in the UK, in practice, are usually laid at depths ranging from 0.75 m in low permeability soils to 1.25 to 1.5m in permeable soils.

2.4 Drain spacing

The required spacing of drains can be calculated using the Hooghoudt equation:

$$L^2 = \frac{8K_b dh}{q} + \frac{4K_a h^2}{q} \tag{2.7}$$

where K_a is hydraulic conductivity above the drain in meters per day, K_b is hydraulic conductivity below the drain in meters per day, h is height of water table above the drain level midway between the drains in meters, q is drain discharge in meters per day and d is equivalent depth -function of depth to impermeable barrier (D) and drain spacing (L).

Values of hydraulic conductivity can be measured in the field using the auger hole method. Alternatively, the designer can use values measured on similar soils elsewhere. The single auger hole method requires a hole some 80 mm in diameter to be bored below the water table. The water in the hole is then pumped or baled out and the rate at which it refills is measured. From these measurements the value of hydraulic conductivity can be calculated.

2.5 Drain flow

Drain pipe sizes can be calculated using the Darcy-Weisbach equation for smooth pipes and Chezy-Manning for corrugated pipes. It is common to 'over design' the pipe to allow for some siltation with the drain capacity normally increased by some 30%. It is normal to assume that the hydraulic gradient line coincides with the pipe soffit, i.e. the pipe flows full.

If the drains are installed in hydraulically unstable soils they will require to be surrounded by a gravel envelope. Generally, soils with a high clay content will be stable and will not require an envelope. Granular envelopes are normally 50 to 100 mm thick. The gradation of the filter should be designed using the US Bureau of Reclamation method.

2.6 Drainage layouts

Collector drains can be open ditches or buried pipes. Buried pipe collectors are to be preferred where sufficient ground slope is available. Pipe drain slopes should not be less than 0.0005 whilst open collector slopes can be as low as 0.0001. To allow drains to be cleaned, they should not exceed 300 m in length without a manhole or outfall into an open channel.

2.7 Surface drainage for irrigated land

Surface drainage is often provided to irrigate land to collect excess irrigation supplies and runoff from rainfall. For surface irrigation typical surface drain capacities can be based on 24 h, 1 in 5yr rainfall with 24 to 48 h storage on the field. For rice drainage, the drain capacity should be sufficient to allow the drawdown of water in the paddies where this is part of the cultivation pattern. Typical values of surface drainage capacity are in the range of 2 to 41/s per hectare.

New Words and Phrase

1.saturation *n.* 可论证的
2.wilt *n.* 茎
3.tension *n.* 终点站
4.deplete *n.* 自然地理学
5.evapotranspiration 前寒武纪地盾
6.prevail 阿巴拉契亚高地
7.incurr *n.* 流域
8.seepage *n.* 地质
9.trickle *n.* 石英岩
10.subirrigation *n.* 变质火山岩
11.topography *n.* 耐久性
12.strip *n.* 陆上风景
13.furrow *n.* 地貌学
14.corrugation *n.* 水文
15.bund *adj.* 湖泊的
16.hectare *adj.* 不减；未变弱
17.tillage *adv.* 尽管如此
18.variant *adj.* 静水的
19.emitter *n.* 动物群
20.Ditch *n.* 沟，渠
21.Germination *n.* 萌发，生长
22.Rot *vi.* 腐烂
23.Poach *vi.* 偷猎
24.Perforate *vi.* 打孔
25.Surplus *n.* 过剩
26.Auger *n.* 木螺钻

Notes

1.With knowledge of the crop rooting depth, the available soil moisture and the crop water requirements, it is possible to select a suitable irrigation interval (time between irrigations).

根据农作物根的深度、可获得土中水含量及农作物对水的需求，我们需要选择恰当的灌溉间隔（每次灌溉的时间间隔）。

soil moisture 土壤湿度 interval 间隔

The transitman sights a rod, which is a rule with spaces marked at regular intervals.

2.It is necessary to account for losses of water incurred during conveyance and application to the field. Efficiency can be divided into three parts: (1) field application; (2) field canal; and (3) distribution efficiency.

必须说明水在传送和应用到土体过程中的损失。有效性可分为三个部分：（1）现场应用；（2）实际管道；（3）分配有效性。

account form 解释，说明

With new scientific discoveries we can now account for many natural phenomena that seemed inexplicable before.

3.Drainage systems can be defined as subsurface and surface. Surface drains are designed to remove excess runoff from the land which would otherwise cause localized flooding.

排水系统可分为表层和地下层。表层排水主要是设计用于排泄超过的径流以免引起当地的洪水。

define 定义 would 从句中指的虚拟语气

Although the processes are presented here as discrete elements with well-defined interfaces, in practice they may overlap and interact in ways not detailed here.

Comprehensive Exercises

I.Answer the following questions on the text.

1.What is the soil moisture?

2.What are the methods for the irrigation?

3.What are the benefits from the drainage of agricultural land?

4.How do we estimate the quantity of water to be removed by a subsurface drainage system?

5.How do we solve the required spacing of drains?

II.Fill the most appropriate words or phrases in the correct forms in the blanks from the list below.

desirable	account for	divided into	topography	flow down
Infiltration	individual	associate with	coincide with	envelope

1.An isolated air supply system is generally used to drive the water out of the runner_____when a turbo-generator is to work as a synchronous compensator.

2.Water is then applied at the head of the_____and allowed to flow down the slope.

3.An economist will normally perform the economic study, closely coordinated with the hydraulics and pumping station design engineers _____the project.

4.In developing countries, buried-pipe distribution systems for surface irrigation about twice the area irrigated by the better known sprinkler and micro-irrigation (trickle) methods.

5.It is normal to assume that the hydraulic gradient line ______the pipe soffit, i.e. the pipe flows full.

6._____is an important process where rainwater soaks into the ground, through the soil and underlying rock layers.

7.______, in large measure, dictates the first choice of type of the dam.

8.Each type of bridge, indeed each______bridge, presents special construction problems.

9.In general, water pollutants ______several broad categories.

10.From a public health or ecological view, a pollutant is any biological, physical, or chemical substance that in identifiable excess is known to be harmful to other _____living organisms.

III.Translate the following sentences into Chinese from the text.

1.Furrow irrigation is used for the irrigation of row crops or crops grown on beds between furrows. Furrow irrigation usually implies sloping land although horizontal furrows can be used for row crops within level basins. Water is applied to the upper end of each furrow and flows down the furrow with water infiltrating into the beds between the furrows on which the crop is grown.

2.Drain pipe sizes can be calculated using the Darcy-Weisbach equation for smooth pipes and Chezy-Manning for corrugated pipes. It is common to 'over design' the pipe to allow for some siltation with the drain capacity normally increased by some 30%.

3.Drainage is essential for agriculture on many of the world's most productive lands. Coastal plains and river deltas may have seasonally or permanently high water tables and must have drainage improvements if they are to be used for agriculture.

4.Various types of irrigation techniques differ in how the water obtained from the source is distributed within the field. In general, the goal is to supply the entire field uniformly with water, so that each plant has the amount of water it needs, neither too much nor too little. Irrigation methods can be divided into four main types-surface, Sprinkler irrigation, localized irrigation and subsurface.

科技英语动词时态及翻译技巧

1. 动词时态用法

科技英语文体的动词形态的用法主要涉及3个方面：

1) 通常采用一般现在时时态

主要因为一般现在时常表述"无时间性"的"一般叙述"，即通常发生和无时限的自然现象、过程、常规等。在科技英语中用一般现在时表述定义、定理、方程式或公式的解释以及图表的说明，描述客观存在事实"无时间性概念"，从而可排除任何与时间牵连的误解，使行文更加生动，文句具有畅晓的说服力。例如：

When the steam touches the cold bottle, it cools and changes into water again. The bottle becomes wet.（水蒸气触到冷瓶子时，就冷却为水，瓶子就湿了。）

The nail is a good conductor of heat .The paper is bad conductor of heat .（钉子是热的良导体，而纸是热的不良导体。）

但在科技英语中有时间状语或告知有时间时，则动词要根据时间的不同而做出相应的时态变化。例如：

Our country has successfully launched some synchronous satellites since 1990.（从1990年起我国已成功发射了几颗同步卫星。）

We had read the operation specifications before we examined the machine.（检修机器前我

们已阅读过使用说明书。)

2) 广泛使用动词的被动语态

由于科技英语叙述的主体多是客观的事物、现象、过程、结论、定律、定义、说明等,而客体往往是从事某种科研工作(包括试验、研究、分析、观测等)的人或设备装置,大量使用动词的被动语态,不仅使论述显得客观,而且可以使读者的注意力集中在叙述中的主体对象上。例如:

(1) 客观叙述,简洁明了

Sometimes the communication would be seriously disturbed by solar spots. (有时候,通信会受到日斑的严重干扰。)

(2) 强调动作的承受者,即主体对象

A rare element was found last year. (去年发现了一种稀有元素。)

(3) 不必或不需指出动作的执行者

Hydrogen is known n to be the lightest element. (大家都知道,氢是最轻的元素。)

3) 广泛使用非限定性动词 (Non-finite Verbs)

即不定式、动名词和分词(包括现在分词和过去分词)。它们在句子中扮演各种成分,从而使科技英语的行文在缺少各种修辞格的情况下保证文章的渲染力,以期增加科学的逻辑性、严密性和普及性。

(1) 动词不定式在科技英语文体中能起到名词、形容词或副词的作用,同时也保留动词的一些特征,在行文中可以作主语、表语、宾语、定语、状语等。

a) 作主语。

To hesitate means failure of operation. (犹豫不决就意味着操作挫败。)

b) 作表语,表示目的、事态发展的结果、同意、安排、命令、决定等。

The factory seems to have been built in the 19th century.

(那个工厂好像是在19世纪建造的。)

c) 作宾语或宾语补足语。

The manufacturer can’t afford to buy the new universal lathe.(制造厂买不起新的万能床。)

d) 作定语,动词不定式与其修饰的词之间往往有动宾关系(如果该不定式是不及物动词,其后应有必要的介词。)

This is good rule to go by. (这是一项应该遵守的好制度。)

(2) 动名词由动词原形“+ing” 构成,兼有动词和名词两者的特征和作用,科技英语中作主语、宾语、表语、定语、状语等。

a) 作主语。

It’s rather tiring testing so many times. (试验这么多次是相当烦人的。)

b) 作表语。

Denying this accident will be shutting one's eyes to fact. (否认这一事故就是闭眼不看事实。)

c) 作宾语。

He imagined finding a new production. (他幻想发现一个新产品。)

d) 作定语。

That's a factory dealing in walking sticks. (那是一家制造手杖的工厂。)

e) 作状语。动名词本身不能作状语，但放在介词后，就可以起状语作用，表示时间、原因、目的、让步、方式等。

He hasn't much experience in running factories. (他没有多少管理工厂的经验。)

(3)分词相对于动词的不定式和动名词来说，在科技英语文章中的应用则更为广泛。它不但具有动词的特点，同时又具有形容词和副词的特征，因而在行文中可以作定语、表语、宾语补足语和状语等。常见的用法有:

a) 作定语。单独的分词用作定语时，可放在所修饰的名词前面，科技英语中常见的是过去分词作定语时，为强调其动作，经常放在所修饰的名词后面。

The fallings tone attracts the earth just as strongly as the earth attracts the stone. (落石对地的吸引力就像地对石头的吸引力一样强。)

b) 作状语。可表示时间、原因、条件、方式或伴随情况等。

A ranging chemical elements according their atomic weights, we find similar ones at certain definite intervals. (如果化学元素按照它们的原子量排列起来，每隔一定数目我们就发现相类似的元素。)

If represented by arrows the forces can be easily computed. (如果用箭头表示，力就可以容易计算。)

而独立主格分词结构，在句中仍用作状语，表示时间、原因或伴随情况。

Other liquids being too light, a barometer uses mercury. (其他液体太轻，因此气压计用水银。)

c) 分词用作宾语补语。一般来说，过去分词用作宾语补语时，表示被动和动作的完成，现在分词用作宾语补语时，表示主动和动作正在进行。

We found the bar magnetized. (我们发现铁棒已磁化。)

The force keeps the body moving . (这个力使该物体继续运动。)

2. 科技英语动词翻译技巧

1) 时态的翻译

科技文章中，英汉两种语言在很多情况下的时态从字面看并不一致。这是由于英汉两种语言的不同特点决定的，但两者表达的概念、条理、逻辑的要求是一致的，词和词项、句子各部分的主题关系也是基本一致的。例如 :Since the middle of the century oil has been in the fore of energy resources , 可译为“ 本世纪石油一直是最重要的能源之一” 。这样英译汉时 , 为了符合汉语的表达习惯 , 就要进行时态转换。科技英语翻译中常见的时态转换有 :

(1) 英语的一般现在时译为汉语的将来时、进行时或过去时。

The cancer reverses completely thanks to early treatment.

(早期治疗癌症是可以完全治愈的。)(将来时)

The electronic computer play an important part in science and techno logy. (电子计算机在科学和技术方面起着重要的作用。)(进行时)

These substances further speed up the decay process. (这些物质进一步加速了衰变过程。)(过去时)

(2) 英语动词的进行时可译成汉语的将来时。

Knowing severe winter is coming would enable squirrel to store plenty of food. (松鼠预知严冬将至而储藏大量的食物。)

(3) 英语动词的完成时翻译成汉语的过去时。

The sales of industrial electronic products have multiplied six times . (工业电子产品销售值增长了 5 倍。)

2) 被动时态的翻译

科技英语文体中大量地使用被动语态 , 主动语态相对较少 , 而汉语则用许多句式来表达被动行为。英译汉时 , 被动语态常可译成汉语的被动语态或主动语态 , 还可译为汉语的无主句或插说。

英语的被动语态译成汉语的被动语态或主动语态。

If light is not completely reflected by the object , some light is said to be absorbed. (如果光未被物体全部反射 , 就说明有些光被吸收了。) (译成被动语态。)

Some metals can be smelted at low temperature. (有些金属可以在低温下冶炼。)(主动语态)

当无须指出动作的执行者的时候 , 英语的被动语态译成汉语的无主句。

3) 科技英语非限定性动词的翻译

作为科技英语中广泛使用的非限定性动词 (分词、不定式和动名词) 在句子中扮演着各种成分 , 英译汉时也要根据它所具有的语法意义和在科技英语中暗含的科技含义转换为适当的汉语词汇意义 , 必要时还要进行词的增补。

(1) 汉语中没有限定动词和非限定动词的分类 , 非限定性动词相当于汉语中的动词 , 而且英语语言主要倚仗形态表意。汉语主要倚仗词汇表意 , 英译汉时为了准确表达出它的科技含义 , 最为常见的是对词的增补和意译。如 :

a) 增补英语中省略的词。

cycling-(周期变化), seeding -(引晶技术), turmeling -(隧道效应)

radio telescope to be used (将要交付使用的射电望远镜)

modulated voltage(已调制电压)

a canning tomato (一种供做罐头的西红柿)

b) 增加关联词语 。

Heated, water will change to vapour.(如水受热 , 就会气化。)

c) 修饰加词 , 语气连贯。

Heat from the sun stirs up the atmosphere , generating winds.(太阳发出的热能搅动大气 , 于是产生了风。)

(2) 当非限定性动词的概念难以用汉语的动词表达 , 翻译时也可转换成汉语的其他词类。

Momentum is defined as the product of the velocity and a quantity called the mass of the body . (动量的定义是速度和物体质量的乘积。)(转换成名词)

There are ten factories of varying sizes in this district . (这个地区有大小十家工厂。)(译成形容词)

UNIT III TYPICAL HYDRAULIC PROJECT

Lesson 1 The Three Gorges Dam

The Yangtze River is the largest and longest river in China, with a drainage area of 1.80 million km^2. The river flows through the Qinghai–Tibet Plateau, Yunnan–Guizhou Plateau, Sichuan Basin, Three Gorges, Jiang-Han Plain, Lower Yangtze Plain, and pours into the East China Sea at Shanghai. The Three Gorges Project is the largest hydro junction project in the world. A key backbone project for the harnessing and exploitation of the Yangtze River, the Three Gorges Project consists of the pivotal dam project, a power transmission and transformation project, and a resettlement project (Fig. 3.1).

Figure 3.1 Locations of the Three Gorges Dam

The axis of the Three Gorges Dam is a total of 2309 m long, the dam crest level is 185 m, and the normal pool level is 175 m. Navigation structures include a permanent navigation lock and a ship lift. The permanent navigation lock is a double-line, five-level, multistage flight lock with a designed annual unidirectional tonnage capacity of 50 million tons. The ship lift is a single-line, one-level, steep-lifting type with a lifting height of 113 m, and a gross ship box weight of about 15500 t the largest lifting height and weight in the world. The power transmission and transformation project involved the construction of 6519 km of 500 kV AC transmission line with a transformation capacity of 22750 MVA and 2965 km of ± 500 kV DC transmission line with a DC converter station capacity of 18000 MW.

After 40 years of planning, surveying, researching, designing, and testing from the early 1950s onwards, the 5th session of the 7th NPC finally approved the Resolution on the Construction of the Yangtze River in April 1992. Construction officially commenced in December 1994, and the diversion closure of the river was carried out in November 1997. In December 2009, the project completed the scheduled preliminary design tasks. The backwater area of the TGR is approximately 660-km long, stretching from Yichang to Chongqing (Fig. 3.1). After the pool level of the TGR was increased to 175 m in 2008, the 432-km long permanent backwater zone stretched from the dam to Fuling, and the fluctuating backwater region extended to Jiangjin (Fig. 3.2). The TGR became fully operational in 2009 and has the largest storage capacity [39.3 billion cubic meters (bcm)] in the Yangtze River basin, constituting approximately 4.5% of the Yangtze's annual discharge.

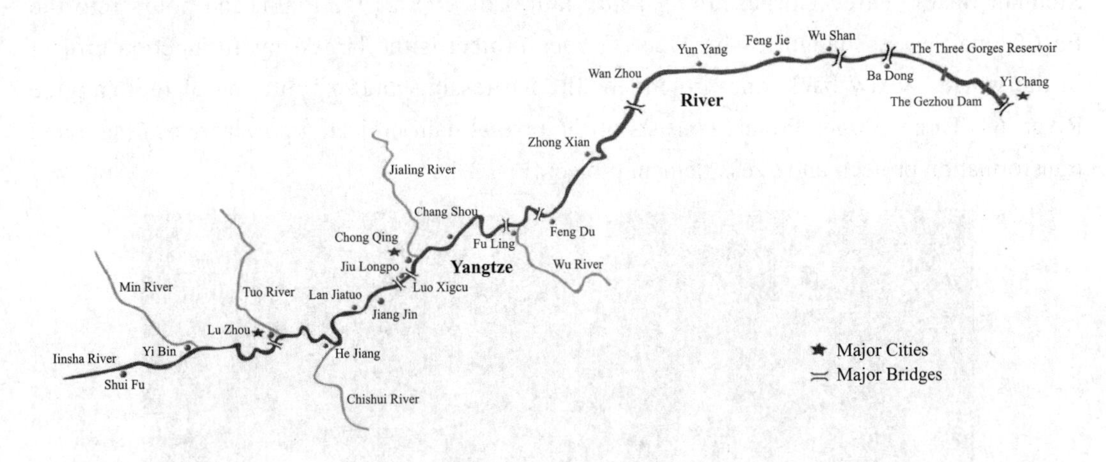

Figure 3.2 Backwater region of the Three Gorges Reservoir

1.Major technical challenges and solutions

The Three Gorges Dam is designed to withstand a 1000-year flood, and to survive a 10000-year flood plus 10%; the corresponding flood discharges are 98800 m^3/s and 124300 m^3/s, respectively. Flood discharge from the dam features a high water head, a large volume of overflow discharge, and a large quantity of sediment outflow. To address both the removal of floating debris in the reservoir, and the diversion required by the construction, a three-layer large outlet layout was adopted after considerable debate, and after modeling with multiple programs. This three-layer outlet required a deep discharge outlet, a surface discharge outlet, and a bottom fluid-diversion outlet in the flood-discharge dam section. It overcame the challenge of dealing with a large volume of flood discharge, and also saved on the engineering work already invested in the project.

The concrete in the Three Gorges Dam occupies as much as 28 million m^3, twice as much as the concrete in the Itaipu Hydroelectric Power Station, and with a shorter construction period. Rapid concrete construction is key to a scheduled performance. The adoption of a tower belt crane led to the new technique of continuous concreting from mixer plant to placement section, realizing the high-intensity feeding of concrete. In the year 2000, the project created a world record of an annual 5.48 million m^3 of concrete placement, with 553500 m^3 per month, and 22000 m^3 per day.

The greatest water depth of the diversion closure was 60 m, and the closure discharge volume was about 8480-11900 m^3/s, with both depth and volume ranking as world records. Pre-flow diversion by a constructed open diversion channel and a pre-dumped bottom cushion reduced the difficulty of the closure, and created a world closure-construction record of 194000 m^3/d of continuous landfill.

The turbo-generator unit of the project has one of the largest unit capacities in the world; the maximum water head variation is 52 m, ranking top among units with similar capacity. The project also developed and adopted a generating unit with "fully air-cooling" and "evaporative-cooling" technologies, providing better operation reliability than "water-cooling" technology. The ship lock of the Three Gorges Project has a designed total water head of 113 m; its delivery head is 45.2 m, and the peak height of the structure is 70 m, making it the highest liner-type ship lock in the world. A high-performance drainage system and a draw-cut type of high-strength bolt structure were adopted to form the world's first "fully liner- type" ship lock; this new lock is a significant and innovative development from the traditional gravity-type lock.

The Three Gorges Project uses a patented double air-cooling technology to realize the stable maintenance of low- temperature concrete at 7℃ at the outlet. The project adopts an individualized dynamic water-cooling technique, and uses a new type of surface insulation material to control the mass concrete temperature and to prevent cracking. More than 4 million m^3 of concrete has been placed for the Phase III dam and, through site inspection, no temperature cracking has been found a miracle in the history of concrete gravity dams.

2.Construction of the Three Gorges Project

Phase I (1994—1997)—The TGP is being constructed in three phases, as shown in Fig. 3.3 Preparations for construction of the dam were done in 1993 and 1994. Formal commencement was declared on Dec. 14, 1994. There was a small island, Zhongbao Island, at the dam axis, which divided the river into two branches. The right branch, Haohe branch, was dry in the low-flow season and floods flowed through it before the project. In the first phase, an earth cofferdam was built to enclose the right branch channel.

Along the right edge of the island a longitudinal dam was built within the cofferdam. The right channel was deepened and widened by excavation to form an open channel capable of conveying a discharge of 70000 m^3/s. The open channel was finished in 1997 and was put into use in May 1997. Ships started to navigate through the open channel in July, 1997. In the meantime, the temporary ship lock was built on the left bank of the main channel. The river was not substantially narrowed and floods passed through the river section smoothly. The permanent ship lock was put into use in this phase.

Phase II (1998—2002)—The beginning of the second phase was marked by the successful closure of the earth coffer dam on the main channel on Nov. 8, 1997. An enclosed area appeared in the river channel and the flood water flowed through the artificial open channel, as shown in Fig.3.3. The main dam was constructed on the river bed within the area enclosed by the coffer dam in the period from 1998—2002. In the fall of 2002 the main dam and the five step ship locks were completed.

Phase III (2003—2009)—In 2002, the cofferdam on the main river channel was removed by explosion and water began to flow through the bottom outlets of the dam. The open channel was cut off by a coffer dam again, as shown in Fig. 3.3. On Nov. 7, 2002, the open channel cofferdam was closed. The new cofferdam was built from elevations of 50-140 m and constructed with rolled concrete. It formed a reservoir in conjunction with the main dam. The water level rose to 135-140 m and the turbo-generators began to generate electricity. The right part of the main dam was built within the cofferdam. Water flowed through the 23 bottom outlets of the spillway section. The permanent ship locks and the ship lift were used for ships to pass through the dam.

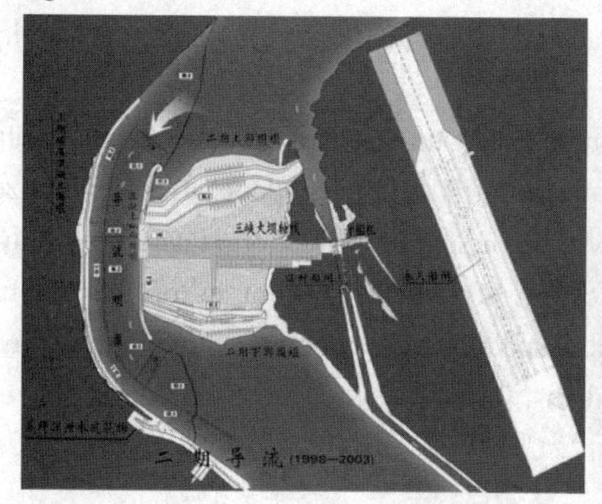

Figure 3.3 Phasing of the TGP construction

3.Major project benefits

After 20 years of hard work, the construction task of the Three Gorges Project is basically complete; the dam is now beginning to produce huge and comprehensive benefits such as flood control, power generation, waterway transport, and water resource utilization.

1) Flood control

The Three Gorges Reservoir has a water area of 1084 km^2 and a storage capacity of 39.3 billion m^3, with a flood-storage capacity of 22.1 billion m^3. On its completion, the Three Gorges Project will improve the flood-control criteria for the Jingjiang river reach from once in a decade to once in a century. During the flood seasons in 2010 and 2012, the Three Gorges Project underwent inflow peak floods of 70000 m^3/s and 71200 m^3/s, respectively, which effectively mitigated the flood-control pressure along the middle and lower reaches of the Yangtze River and protected the lives and property of over 15 million people and 23 million mu of land.

2) Power generation

The Three Gorges power station has 34 power sets (32 units with a unit capacity of 700 000 kW and 2 units with a unit capacity of 50000 kW) with an installed gross capacity of 22.5 million kW. Its annual power generation is 100 billion kW · h, which is equivalent to a yearly reduction of 50 million tons of raw coal consumption and a yearly reduction of 100 million tons in carbon dioxide emissions. In 2014, the Three Gorges power station generated a total of 98.8 billion kWh of electricity, exceeding the world record for annual power generation from a single hydroelectric power station, a record that had previously been created by the Itaipu Binacional. With a total electricity generation exceeding 800 billion kW · h by the end of 2014, the Three Gorges power station has effectively relieved the shortage of electricity in East China, Central China, and Guangdong, and has become a valuable large production base of clean energy in China.

3) Navigation

The Three Gorges Project improves the navigation conditions from Yichang to Chongqing, a distance of about 660 km, and along the middle and lower reaches of the Yangtze River in the low-flow season. As a result, a 10000 tonner fleet can travel all the way from Hankou to Chongqing, reducing transportation costs significantly. In 2011, the freight volume through the Three Gorges navigation lock reached 100.3 million tons, achieving its planned target 19 years in advance. Over the lock's 10 years of navigation, the accumulated freight in transit has reached 700 million tons. Thus, the Yangtze River has become an extremely profitable watercourse.

4) Comprehensive water resource utilization

The abundant fresh water stored in the Three Gorges Reservoir is a valuable strategic freshwater resource in China, and could provide a guaranteed water source for water deficient areas in northern China. The Three Gorges Reservoir could supplement the middle and lower reaches of the Yangtze River with more than 20 billion m^3 in the low-flow season every year, adding an average depth of about 0.8 m to the main channel. This supplementary water could effectively mitigate shortages in water for production, living, and ecological needs along the middle and lower reaches of the Yangtze River.

In spite of the stunning benefits brought by the Three Gorges Project, it will also cause certain impacts on peripheral areas and the downstream eco-environment. In particular, the reservoir shore reconstruction might give rise to geological hazards and have impacts on the water quality and aquatic life in the reservoir, as well as in the downstream rivers and lakes. Therefore, we must pay close attention to these matters.

4.Future work

To fit in with the new strategic positioning of the Three Gorges Project and the reservoir region in coming years, the Chinese government has made a significant deployment of resources toward subsequent work in the Three Gorges region. With the wellbeing and enrichment of millions of migrants as the intent and purpose of subsequent work, we should accelerate the development of a generally affluent society in the reservoir region. We should give priority to the reestablishment and protection of the ecological environment; ensure the effective protection of this national strategic freshwater resource pool; and strengthen our observation and treatment of impounding influences, making use of advantages while striving to avoid disadvantages. We should also enhance our integrated management, optimize our regulation, and improve our ability in and level of scientific management for the Three Gorges Project. In this way, we can ensure the long and safe running of the dam and sustain its comprehensive benefits; drive economic and social development in the Three Gorges region, enabling the development of a harmonious society; improve the ability of the Three Gorges Project to serve the national economy and nationwide social development; and thus provide improved benefits to the people.

New Words and Phrase

1.backbone *n.* 支柱，脊梁
2.harness *vi.* 利用，控制
3.exploitation *n.* 开发
4.resettlement *n.* 安置
5.lock *n.* 船闸
6.ship lift 升船机
7.transmission line 传输线
8.backwater *n.* 回水
9.water head 水头
10.diversion *n.* 分流
11.tower belt crane 塔带机
12.cushion *n.* 垫
13.turbo-generator *n.* 涡流发电机
14.insulation *n.* 隔离，孤立
15.drainage *n.* 排水
16.cofferdam *n.* 围堰
17.utilization *n.* 使用，利用
18.mitigate *vi.* 减轻
19.dioxide *n.* 二氧化碳
20.hydroelectric power station 水电站
21.deficient *adj.* 不足的，缺乏的
22.freight *n.* 货运，货物
23.valuable *adj.* 有价值的
24.profitable *adj.* 有意义的
25.peripheral *adj.* 周边的
26.wellbeing *n.* 健康，幸福
27.strive *vi.* 努力奋斗，斗争
28.harmonious *adj.* 和谐的
29.deployment *n.* 部署

Notes

1.The river flows through the Qinghai-Tibet Plateau, Yunnan-Guizhou Plateau, Sichuan Basin, Three Gorges, Jiang-Han Plain, Lower Yangtze Plain, and pours into the East China Sea at Shanghai.

长江流经青海高原、云贵高原、四川盆地、三峡以及江汉平原、长江下游平原、并

在上海汇入东海。

low through 流经 in respect of 关于，就……而言，比较

During the second period, the river water will flow through the diversion channel and shipping will use either the open diversion channel or the temporary ship lock.

similar to 注入；灌注；倾注

Thousands of tons of filth and sewage pour into the Ganges every day.

2.After the pool level of the TGR was increased to 175 m in 2008, the 432-km long permanent backwater zone stretched from the dam to Fuling, and the fluctuating backwater region extended to Jiangjin.

当 2008 年三峡坝前水位达到 175m 时，库区常年回水区为坝前到涪陵，共 432km, 变动回水区延长至江津。

pool level 坝前水位 stretch from to 从……延伸到

Its consequences stretch from the price of steel to the development of a multibillion - dollar derivatives market.

3.To fit in with the new strategic positioning of the Three Gorges Project and the reservoir region in coming years, the Chinese government has made a significant deployment of resources toward subsequent work in the Three Gorges region.

为适应未来三峡工程及库区战略定位，中国政府对三峡区域的后续资源利用进行了重要部署。

fit in with 适合 ; 使与……一致 ; 与……一致 ; 与……相适应

It is difficult to make hardcut military negotiations fit in with flexible diplomacy.

Comprehensive Exercises

I.Answer the following questions on the text.

1.How many province does the Yangtze River flow through?

2.What purpose does the Three Gorges Project serve?

3.What are the main technical challenges in the TGP construction? How to solve them?

4.How long is the backwater region of TGR?

5.What is the future work for the Three Gorges Project?

II.Fill the most appropriate words or phrases in the correct forms in the blanks from the list below.

pour into	harness	discharge	diversion	put into use
cutoff	valuable	supplementary	make use of	optimize

1.The Three Gorges Dam is the key to______and exploit the Yangtze River, and the main Functions of this huge project are Flood control, power generation, improved navigation etc.

2.How can a proper balance be achieved between this use and other ______, and often conflicting, uses, such as municipal water supply and recreation?

3.The system_____measure the velocity and water level during the flooding-season.

4.Annual mean precipitation is 648 mm in china, with an estimated ________of 2,711.5 billion m^3.

5.The meandering river_____when the sinuosity exceeds the critical value

6.The team_____all the resources _____the project.

7.Some areas recycle tires by using them as _____fuel in cement kilns.

8.The money estimates need to be placed on a comparable basis by appropriate conversions that______the mathematics of compound interest.

9.Google promises further mitigations and memory _____in future Chrome updates.

10.Canal intakes are located a short distance upstream from the_____structure and serve to regulate the flow into the channel.

III.Translate the following sentences into Chinese from the text.

1.In the TGR, the 660 km long waterway from Yichang to Chongqing will be improved, making it possible for 10000 tons of shipping to sail directly upstream to Chongqing. The river transport will be improved from 10 million to 50 million tons per year and the costs reduced by 35%-37%. Shipping will become safer although the gorges have been notoriously dangerous to navigate.

2.The Three Gorges Dam is being built at Sandouping in the middle of central China, and the site is located in the middle section of the Xiling gorge on the Yangtze River. The project designed total storage capacity of the reservoir is 39.3 billion m^3, and will create a 58000 km^2 Three Gorges Reservoir Area.

3.With a total electricity generation exceeding 800 billion kWh by the end of 2014, the Three Gorges power station has effectively relieved the shortage of electricity in East China, Central China, and Guangdong, and has become a valuable large production base of clean energy in China.

4.In this way, we can ensure the long and safe running of the dam and sustain its comprehensive benefits; drive economic and social development in the Three Gorges region, enabling the development of a harmonious society; improve the ability of the Three Gorges Project to serve the national economy and nationwide social development; and thus provide improved benefits to the people.

Reading Material Reservoir Sedimentation

A wide range of sedimentation related problems occur upstream of dams as a result of sediment trapping. Because of storage loss the functions of the reservoir reduce for flood control, power generation, and water supply. Sediment can enter and obstruct intakes and greatly accelerate abrasion of hydraulic machinery, thereby decreasing its efficiency and increasing maintenance costs. Sediment deposition in the delta region in the reservoir may affect navigation and impact the ecology. Dam construction is the largest single factor influencing sediment delivery to the downstream reaches. The cutoff of sediment transport by the dam can cause stream bed degradation, accelerate the rates of bank failure, and increase scour at structures such as bridge piers.

1.Patterns of reservoir sedimentation

Patterns of reservoir sedimentation depend on the operational scheme of the dam, hydrologic conditions, sediment grain size, and reservoir geometry. In reservoirs with fluctuating water levels or that are periodically emptied, previously deposited sediments may be extensively eroded by processes such as downcutting by stream flow. Further complexity is added when there are significant sediment inputs from tributaries. Most sediment is transported within reservoirs to points of deposition by three process:

(1) Transport of coarse material as bed load along the topset delta deposits; (2) transport of fine sediment in turbid density currents; and (3) transport of fine sediment as suspended load.

The longitudinal sediment deposition exhibits five basic types depending on the inflowing sediment characteristics and reservoir operation, as shown in Fig. 3.4.

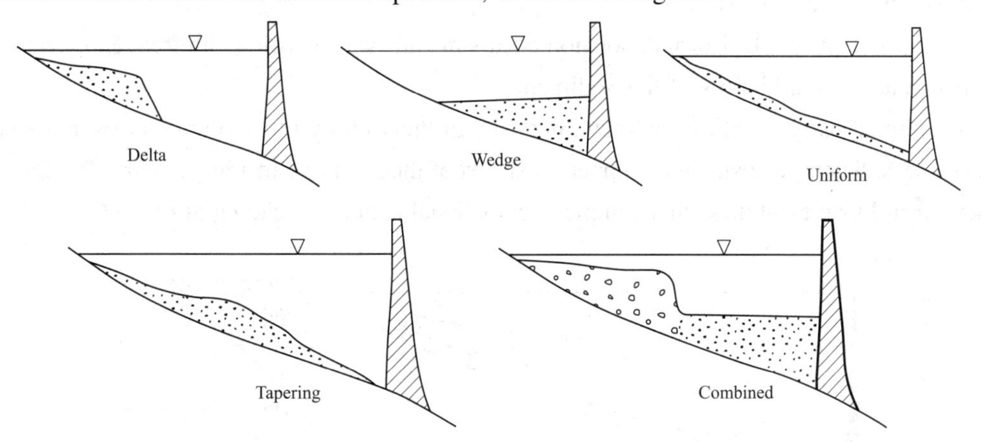

Figure 3.4 Longitudinal patterns of sediment deposition in a reservoir

Delta—A delta shaped deposit is formed with the coarse fraction of the sediment load, which is rapidly deposited at the upper zone of the reservoir. It may consist of coarse sediment or may also contain a fraction of finer sediment such as silt. Such a deposition pattern occurs if the reservoir remains at high pool level for a long time.

Wedge—A wedge shaped deposit is a typical pattern of fine sediment deposition by turbidity currents. It occurs in small reservoirs with a large inflow of fine sediment and in large reservoirs operated at a low pool level during flood events, which causes most sediment to be carried to the vicinity of the dam. Figure 3.5 shows the profile of the Bajiazui Reservoir on the Puhe River as an example of the wedge shaped deposit.

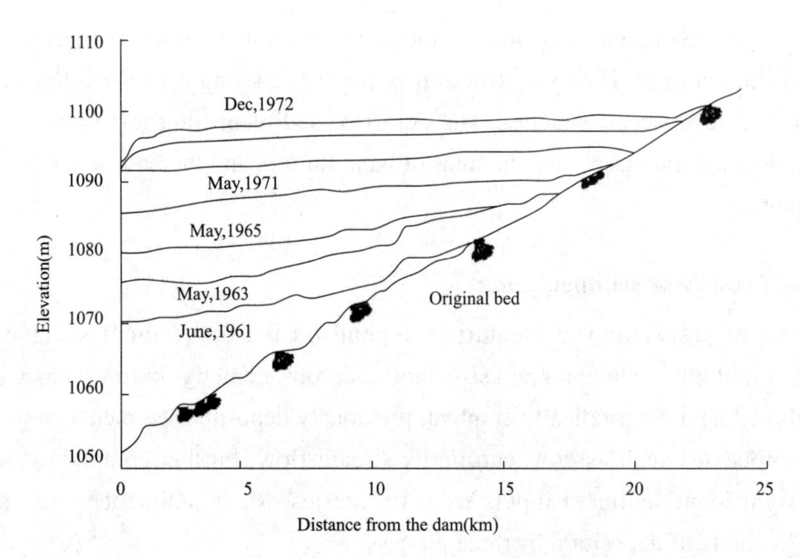

Figure 3.5 Profile of the Bajiazui Reservoir on the Puhe River

Tapering deposit—A tapering deposit occurs in large reservoirs normally remaining at a high pool level, it is formed due to the progressive deposition of fine sediment from the flowing water moving toward the dam.

Uniform deposit—Uniform deposition occurs in narrow reservoirs with frequent water level fluctuation and a small inflow of fine sediment.

Combined patterns—Fine sediment deposits in the vicinity of the dam and form a wedge and coarse sediment deposits in the upper most part of the reservoir forming a delta. The Sakuma Reservoir in Japan exhibits such a complex depositional pattern, as shown in Fig. 3.6.

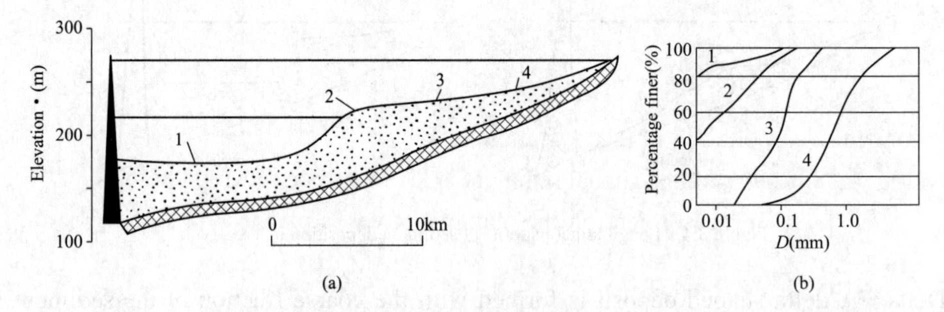

Figure 3.6 Profile of the Sakuma Reservoir in Japan

(a) Complex longitudinal profile of the Sakuma Reservoir in Japan; (b) Grain size distributions of deposits at different sites in the reservoir

2.Sedimentation management strategies

1) Empty flushing

There are two general categories for flushing: (1) empty or free flow flushing, which involves emptying the reservoir to the level of the flushing outlet with riverine flow through the impoundment, and (2) drawdown or pressure flushing, which requires less drawdown but is also less effective. The second method is not commonly used. Empty flushing can also be classified according to whether it occurs during the flood season or the non-flood season. While both strategies have been applied successfully, flood season flushing is generally more effective because it provides larger discharges with more erosive energy, and flood borne sediments may be routed through the impoundment. Flushing scours a single main channel through reservoir bed while floodplain deposits on either side are unaffected. Profile views of a flushing channel are shown in Fig. 3.7.

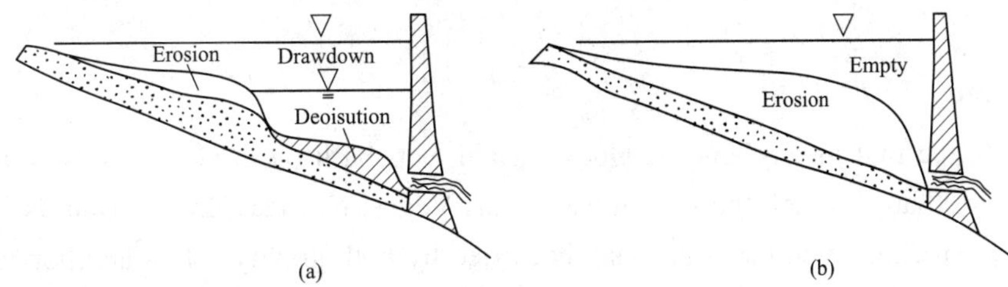

Figure3.7 Profile of a flushing channel

(a) Drawdown flushing causes erosion in the upper part of the reservoir and re-deposition near the dam, with pressure flow through the bottom outlets; (b) Empty flushing results in erosion in the whole reservoir

2) Pressure flushing

Hydraulic flushing involves reservoir draw-down by opening a low level outlet to temporarily establish riverine flow along the impound reach, flushing the eroded sediment through the outlet. Sediment entering the reservoir during flushing periods is also released. In general a conical scour hole in front of the outlet is formed during the flushing. Sediment from the upper portion of the reservoir is transported towards the dam during draw down, but only material in the scour hole can be flushed out.

3) Storing the clear and releasing the turbid

Sediment transportation in many Chinese rivers occurs mainly in 2-4 months of the flood season, that is, 80%-90% of the annual sediment load is transported with 50%-60% of the annual runoff. The Three Gorges Project on the Yangtze River is planned for flood control, power generation, and inland navigation. For these purposes, it is important to maintain adequate storage in the reservoir. The main strategy to control sedimentation is to draw down the pool level from 175 m to 145 m in the flood season from June to September when the sediment concentration is high and allow the turbid water to wash downstream through the reservoir. The reservoir starts to store water in October when the inflowing water becomes clear (i.e. has a lower sediment concentration).

4) Dredging

Dredging has been used for a long time for sedimentation management in small reservoirs. Different dredging methods are applied in reservoir management, including mechanical dredging and dumping outside of the reservoir and agitating sediment with jets so that that sediment can be transported downstream of the reservoir by currents. Various dredgers have been used: dredge boat; agitating dredger; dipper dredger; hauling scraper; excavator and bulldozer; and trailer dredger. In general dredging is more expensive than other strategies of sedimentation control in reservoirs. Nevertheless, jets in combination with turbidity currents or flushing are more economically feasible and have been applied in large reservoirs.

Lesson 2 Hoover Dam and Aswan Dam

1.Hoover dam

In 1935, the first mega-dam, the Hoover Dam, was built on the Colorado River in Black Canyon, near what was then the little town of Las Vegas, Nevada. Hoover Dam is a concrete arch-gravity type, in which the water load is carried by both gravity action and horizontal arch action. The first concrete for the dam was placed on June 6, 1933, and the last concrete was placed in the dam on May 29, 1935.

Standing 726 feet above bedrock, three times the size of the Statue of Liberty, it is still the Western Hemisphere's highest concrete dam. It is 660 feet thick at its base, 45 feet thick at its crest, and stretches 1244 feet across the Black Canyon. Hoover Dam contains three and one-quarter million cubic yards of concrete. There are 4360000 cubic yards of concrete in the dam, power plant, and appurtenant works.

1.1 The reservoir

The reservoir, Lake Mead, is America's largest man-made reservoir. At elevation 1221.4, Lake Mead contains 28537000 acre-feet. Hoover Dam's massive concrete walls held back the waters of Lake Mead, and the reservoir extends approximately 110 miles upstream toward the Grand Canyon.

The width varies from several hundred feet in the canyons to a maximum of 8 miles .The reservoir covers about 157900 acres or 247 square miles.

1.2 The power plant

There are 17 main turbines in Hoover Power plant .The original turbines were all replaced through an uprating program between 1986 and 1993. The plant has a nameplate capacity of 2074000 kilowatts. This includes the two station-service units, which are rated at 2400 kilowatts each. From 1939 to 1949, Hoover Power plant was the world's largest hydroelectric nameplate: with an installed capacity of 2.08 million kilowatts, it is still one of the country's largest.

1.3 Finance

Hoover Dam's $165 million cost has been repaid, with interest, to the Federal Treasury through the sale of its power. Hoover Dam energy is marketed by the Western Area Power Administration to 15 entities in Arizona, California, and Nevada under contracts which expire in

2017. Most of this power, 56 %, goes to southern California users; Arizona contractors receive 19 %, and Nevada users get 25 % .The revenues from the sale of this power now pay for the dam's operation and maintenance.

1.4 Multipurpose benefits

Hoover Dam pioneered the Bureau of Reclamation's efforts in multiple-purpose water resources development.

Hoover Dam's reservoir, Lake Mead, can store nearly 2 years of average Colorado river flow .This water is released in a regulated, year-round flow as needed. Colorado River water irrigates more than a million acres of land in this country and nearly one-half million acres in Mexico. The water helps meet the municipal and industrial needs of over 14 million people.

As it passes through Hoover's turbines, the water generates low-cost hydroelectric power for use in Nevada, Arizona, and California. Hoover Dam alone generates more than 4 billion kilowatt-hours a year-enough to serve 1.3 million people.

Water that was once muddy is now sparkling clear in reservoirs and in stretches of the river. Hoover and other dams on the Colorado have tamed the turbulent flow, creating clean bodies of water that provide recreation. These waters have also formed habitats for fish and wildlife in lands that were once nearly barren.

Lake Mead is one of America's most popular recreation areas, with a 12-month season that attracts more than 9 million visitors each year for swimming, boating, skiing, and fishing.

Hoover Dam changed the Colorado River from a natural menace to a national resource. Hoover and other dams on the Colorado diverted the Colorado River's water into Arizona, Nevada, and southern California, fueling the growth of major cities and helping to turn the arid West into a lush and lucrative garden.

2.Aswan high dam

Aswan is a city on the first cataract of the Nile in Egypt. Two dams straddle the river at this point: the newer Aswan High Dam, and the older Aswan Dam or Aswan Low Dam. Without impoundment the River Nile would flood each year during summer, as waters from East Africa flowed down the river. These floods brought nutrients and minerals that made the soil around the Nile fertile and ideal for farming. As the population along the river grew, there came a need to control the flood waters to protect farmland and cotton fields. In a high-water year, the whole crop may be entirely wiped out, while in a low-water year there was widespread drought and famine. The aim of this water project was to prevent the river's flooding, generate electricity and provide water for agriculture.

2.1 Construction history

The British began construction of the first dam in 1899. Construction lasted until 1902. It was opened in December 10, 1902. The project was designed by Sir William Willcocks and involved several eminent engineers including Sir Benjamin Baker and Sir John Aird, whose firm,

John Aird & Company, was the main contractor. A gravity dam was 1900 m long and 54 m high. The initial design was soon found to be inadequate and the height of the dam was raised in two phases, 1907–1912 and 1929–1933.

When the dam almost overflowed in 1946 it was decided that rather than raise the dam a third time, a second dam would be built 6 km upriver (about 4 miles). Proper planning began in 1952, just after the Nasser revolution, and at first the USA and Britain were to help finance construction with a loan of USD $270 million. Both nations canceled the offer in July 1956 for reasons not entirely known. A secret Egyptian arms agreement with Czechoslovakia (Eastern Bloc) and Egyptian recognition of the People's Republic of China are cited as possible reasons. Soon thereafter, Nasser nationalized the Suez Canal, intending to use its tolls to subsidize the High Dam project. This prompted Britain, France, and Israel to attack Egypt, occupying the Suez Canal and precipitating the Suez Crisis. The United Nations, USSR and US forced the invaders to withdraw and the canal was left in Egyptian hands. The Egyptian government continued to intend to finance the dam project alone by using the revenues of the Suez Canal to help pay for construction. But as part of the Cold War struggle for influence in Africa the Soviet Union stepped in 1958, and they provided technicians and heavy machinery. The enormous rock and clay dam was designed by the Russian Zuk Hydroproject Institute.

Construction began in 1960. The High Dam was completed on July 21, 1970, with the first stage finished in 1964. The reservoir began filling in 1964 while the dam was still under construction and first reached capacity in 1976.

2.2 Benefits

The Aswan High Dam is 3600 m in length, 980 m wide at the base, 40 m wide at the crest and 111 m tall. It contains 43 million m³ of material. At maximum, 11000 m³ of water can pass through the dam every second. There are further emergency spillways for an extra 5000 m³ per second and the Toshka Canal links the reservoir to the Toshka Depression. The reservoir, named Lake Nasser, is 550 km long and 35 km at its widest with a surface area of 5250 km² and holds 132 km³.

The dam powers twelve generators each rated at 175 megawatts, producing a hydroelectric output of 2.1 gigawatts. Power generation began in 1967. When the dam first reached peak output it produced around half of Egypt's entire electricity production (about 15% by 1998) and allowed for the connection of most Egyptian villages to electricity for the first time. The dam mitigated the effects of dangerous floods in 1964 and 1973 and of threatening droughts in 1972–73 and 1983–84. A new fishing industry has been created around Lake Nasser, though it is struggling due to its distance from any significant markets.

2.3 Environmental and cultural issues

In addition to the benefits, however, damming the Nile caused a number of environmental issues. It flooded much of lower Nubia and over 90000 people were displaced. Lake Nasser flooded valuable archeological sites. The silt which was deposited in the yearly floods, and made the Nile floodplain fertile, is now held behind the dam. Silt deposited in the reservoir is lowering the water storage capacity of Lake Nasser. Poor irrigation practices are waterlogging soils and bringing salt to the surface. Mediterranean fishing declined after the dam was finished because nutrients that used to flow down the Nile to the Mediterranean were trapped behind the dam.

There is some erosion of farmland down-river. Erosion of coastline barriers, due to lack of new sediments from floods, will eventually cause loss of the brackish water lake fishery that is currently the largest source of fish for Egypt, and the subsidence of the Nile Delta will lead to inundation of the northern portion of the delta with seawater, in areas which are now used for rice crops. The delta itself, no longer renewed by Nile silt, has lost much of its fertility. The red-brick construction industry, which used delta mud, is also severely affected. There is significant erosion of coastlines (due to lack of sand, which was once brought by the Nile) all along the eastern Mediterranean.

The need to use artificial fertilizers supplied by international corporations is controversial too, causing chemical pollution which the traditional river silt did not. Indifferent irrigation control has also caused some farmland to be damaged by waterlogging and increased salinity, a problem complicated by the reduced flow of the river, which allows salt water to encroach further into the delta.

2.4 A wall commemorating the completion of Aswan High Dam

Mediterranean fish stocks are also negatively affected by the dam. The eastern basin of the Mediterranean is low in fertility, and traditionally the marine ecosystem depended on the rich flow of phosphate and silicates from the Nile outflow. Mediterranean catches decreased by almost half after the dam was constructed, but appear to be recovering. The dam has been implicated in a rise in cases of schistosomiasis (bilharzia), due to the thick plant life that has grown up in Lake Nasser, which hosts the snails who carry the disease.

The Aswan dam tends to increase the salinity of the Mediterranean Sea, and this affects the Mediterranean's outflow current into the Atlantic Ocean. This current can be traced thousands of kilometers into the Atlantic. Some people believe that the dam's effect on this outflow speeds up processes that lead to the next ice age.

New Words and Phrase

1.stretch *vi.* 拉长；拽宽；撑大
2.appurtenant *adj.* 从属的，附属的
3.canyon *n.* 峡谷
4.turbine *n.* 水轮机
5.revenue *n.* 收益，财政收入
6.municipal *adj.* 市政的
7.sparkle *vi.* 闪烁，闪耀，生气勃勃
8.barren *adj.* 贫瘠的
9.menace *n.* 威胁；危险的人
10.lucrative *adj.* 获利多的
11.impoundment *n.* 储水量；围住
12.archeological *adj.* 考古的

13.waterlogging *n.* 水涝；淹水
14.decline *vi.* 减少，降低
15.controversial *adj.* 有争议的
16.encroach *vi.* 侵占

Notes

1.Hoover Dam is a concrete arch-gravity type, in which the water load is carried by both gravity action and horizontal arch action.

胡湖大坝是混凝土拱形重力坝，该坝型的水荷载主要通过重力作用和水平拱推力承担。

water load 水荷载

该句型为非限制性定语从句，in which 相当于 where。

Another type of bridge is the cantilever, in which a horizontal beam extends beyond its support.

2.In a high-water year, the whole crop may be entirely wiped out, while in a low-water year there was widespread drought and famine.

丰水年期间，整个庄稼将会被完全摧毁，而在枯水年则会是大面积的干旱和饥荒。

wipe out 摧毁 widespread 普遍的，广泛的

The deaths occurred when police acted to stop widespread looting and vandalism.

3.Erosion of coastline barriers, due to lack of new sediments from floods, will eventually cause loss of the brackish water lake fishery that is currently the largest source of fish for Egypt, and the subsidence of the Nile Delta will lead to inundation of the northern portion of the delta with seawater, in areas which are now used for rice crops.

由于缺少洪水带来的泥沙补给，海岸线屏障的侵蚀将最终带来咸水湖泊的减少，而这些湖泊是埃及鱼类主要来源，同时尼罗河三角洲的下沉将引起三角洲北部地区长时间淹没于海水中，而这些地区又是大米的主要产地。

be used for 用于 subsidence 下沉

The electrical power, which is consequently obtained at the terminals of the generator, is then transited to the area where it is to be used for doing work.

Comprehensive Exercises

I.Answer the following questions on the text.

1.What are the type of the Hoover dam and Aswan dam?

2.What are the main benefit from the Hoover dam?

3.What can we obtain from the construction of the dam?

4.What are the influence on the environment for the Aswan dam?

5.How can we solve the problem that result from the water engineering projects?

II.Fill the most appropriate words or phrases in the correct forms in the blanks from the list below.

vary	pass through	wipe out	rather than	promote
deposit	erosion	depend on	implicate	lead to

1.When phosphates or nitrates are added to the surface water, they act as a fertilizer and____the growth of undesirable algae populations.

2.These contributions, however, _____the standard of survey, design and construction, and the limited knowledge of buried-pipe distribution systems among irrigation engineers is an important constraint.

3.Compare with the gravity dam, arch dams transfer the greater proportion of the water load to the valley sides ______to the floor.

4.Another important factor in this situation is logging practices that increase soil erosion, _______sediment pollution in streams.

5.These particles are simply too large to_______the pore spaces of the medium and become trapped in the upper depths of the filter.

6.The terminal structure prevents excessive________of the stream channel or damage to adjacent structures and the dam from the high-energy spillway discharges.

7.For heads_____between 15 m to 90 m, reservoir pump turbines have been devised, which can function both as a turbine as well as a pump.

8.Since then other metals have been_______as causative agents in nervous disorders, bone weakening, and heart disease.

9.Erosion of the side slopes______material in the spillway channel, especially where the side slopes meet the channel bottom.

10.In a high-water year, the whole crop may be entirely______by the flood.

III.Translate the following sentences into Chinese from the text.

1.Water that was once muddy is now sparkling clear in reservoirs and in stretches of the river. Hoover and other dams on the Colorado have tamed the turbulent flow, creating clean bodies of water that provide recreation. These waters have also formed habitats for fish and wildlife in lands that were once nearly barren.

2.Hoover Dam changed the Colorado River from a natural menace to a national resource. Hoover and other dams on the Colorado diverted the Colorado River's water into Arizona, Nevada, and southern California, fueling the growth of major cities and helping to turn the arid West into a lush and lucrative garden.

3.The need to use artificial fertilizers supplied by international corporations is controversial too, causing chemical pollution which the traditional river silt did not. Indifferent irrigation control has also caused some farmland to be damaged by waterlogging and increased salinity, a problem complicated by the reduced flow of the river, which allows salt water to encroach further into the delta.

4.Mediterranean fish stocks are also negatively affected by the dam. The eastern basin of the Mediterranean is low in fertility, and traditionally the marine ecosystem depended on the rich flow of phosphate and silicates from the Nile outflow. Mediterranean catches decreased by almost half after the dam was constructed, but appear to be recovering.

Reading Material Dams and Canals

Among mankind's oldest works are irrigation and water supply systems; indeed, the earliest civilizations in river valleys in the Middle East were based on agriculture that depended on irrigation. Harbor facilities and canals for navigation were also early engineering accomplishments. Such systems are usually grouped together as hydraulic engineering projects. Hydraulics is the science that deals with the flow and control of water and other fluids.

Among the most impressive modern works in hydraulic engineering are such great dams as the Aswan High dam on the Nile in Egypt and the Hoover dam on the Colorado River in the southwestern United State. These dams, like most modern dams sever a number of different purposes. Among them are flood control, water storage, irrigation, navigation, and hydroelectric power (generating electricity by means of waterpower). In addition, most dams are also links in a highway system, with a roadway running across them, the lakes behind them, like Lake Mead which is backed up by the Hoover dam, also sever as recreational areas.

Before design and construction of a dam can begin, an extensive survey and study of the site must be made, this survey examines not only topographical features of the area, but also soil and rock samples to determine the geological factors that may affect it. The hydraulic features of the stream or river that is being dammed must also be determined the rate of volume of flow of the river at different dam. Engineers use this information to calculate potential water pressure. It is also necessary to study the site to see whether the dam can be constructed with the use of cofferdams or whether the flow of the river must be diverted. Cofferdams, as we mentioned previously, are watertight piles that form an enclosure from which water can be pumped. When it is necessary to divert the river, one technique is to dig tunnels for the channel; another is to excavate a temporary channel for the river around the dam site.

Even after the site has been made. The preliminary work is still not complete. Scale models of the dam are often made so that they can be tested under simulated conditions. Computers are also used extensively to calculate all the different stresses, to which such huge structures can be subjected, including those that may be caused by earthquakes.

Basically there are two types of construction for dams. Masonry and embankment before the invention of Portland cement, huge blocks of cut stone were ordinarily used to build dams, but today masonry dams are constructed with reinforced concrete. Masonry dams are most often built to control swift—flowing streams in narrow valleys where there is good rock for the foundations. An excellent example is the Hoover dam. Embankment dams are essentially great mounds of earth across a stream. In addition to compacted earth, embankment dams can be built with crushed rock or sand. These kinds of dam are usually built across wider streams where the water flows rather slowly. The Aswan High dam is a good example of an embankment dam.

The velocity and pressure of the water that is being blocked are important factors in the design of dams. Another factor is the possibility of seepage under the foundations, often requiring special protective features in the design. Seepage is the slow leaking of water through a porous material, such as earth or some kinds of rock like limestone or sandstone.

Masonry construction is used for the type of dam known as a gravity dam. The weight of the dam counteracts the force of water pressure. Gravity dams have a triangular profile, with the vertical side facing upstream. The triangular shape is necessary because water pressure increases with depth, so the dam must opposite greater thrust at the bottom. Some masonry dams combine the gravity type of design with an arch or semicircular shape. The pressure of design with a helps to keep the joints between the blocks of masonry closed, so less material is needed to construct a gravity-buttresses; in such cases, the wall facing upstream is usually inclined at an angle.

Embankment dams are great mounds of compacted earth, crushed rock, or sand. The two most common types are earth—fill dams and rock—fill dams. They are angled on both upstream and downstream sides, as though two right triangles had been placed back-to-back. In modern embankment dams it is common practice to provide a core built with some material that water couldn't penetrate. In the Aswan High there is a clay core that rests on a foundation of grouted clay and sand. Beneath the foundations there is a curtain of grout that extends to a deep of 982 meters beneath the riverbed. Other embankment dams also have various zones where special materials strengthen the structure against the danger of seepage.

Many dams have other auxiliary structures, opening on the reason why the dam was constructed. One feature is a spillway that allows floodwater or excess water from the reservoir behind the dam to be released downstream. With embankment dams, the spillways are ordinarily constructed at one side of the dam. With concrete gravity dams, the sloping downstream face often acts as the spillways. In this case some kind of footing or special device must be placed at the bottom of the dam so that the water is projected out into the stream where it cannot erode the dam's foundation.

Other opening is necessary when the dam is used for irrigation or for generating electricity. Gates are built in the dam through which water can be released for these purposes. The gates are equipped with screens so that floating objects cannot pass through them. The ducts that carry water from the gates to turn turbines in a powerhouse are called penstocks. Some dams also have fish ladders that allow fish in the river to travel past the dam to or from their breeding grounds.

Among the most famous accomplishments of modern hydraulic engineering are the three great international canals. The Kiel Canal in Germany connects the Baltic and North Seas, a distance of 95 kilometers .The Suez Canal in Egypt connects the Mediterranean Sea with the Red Sea, providing a passageway between Europe and Asia that eliminates the voyage around Africa. It is 169 kilometers long and has no locks. Locks are sections of a canal that are enclosed by gates; the level of water within the lock can be regulated so that shipping can be raised or lowered to different elevations. The Suez Canal was originally opened in 1860, closed in 1967 because of warfare in the area, and reopened in 1975. Engineers currently plan to widen and deepen the Suez Canal to accommodate the supertankers that carry oil from the oil fields on the Persian Gulf.

Ferdinand de Lesseps, the Frenchman who designed and superintended the building of the Suez Canal, also tried to build a canal across the Isthmus of Panama to connect the Atlantic and Pacific Oceans. He was defeated by the rugged terrain and by the climate in which mosquitoes carried yellow fever. The United States took over the construction of the canal, but it was not until yellow fever had been eliminated and design problems overcome that construction could begin. The canal was finally opened in 1914. It covers a distance of 85 kilometers and has two principal sets of locks.

Impressive and important as this three canals are, they represent not only a small proportion of the world's canals or canalized rivers, that is, rivers that have been dredged, straightened, embanked, or otherwise controlled so that they are navigable. Both the Rome and the Chinese in ancient times built canals .The Grand Canal in China that connects the Yangtze River with Peking is one of greatest engineering works in any age; it is still used today. In Europe a new age of canal—building began in the seventeenth century long before railroads began to spread across the landscape .An important canal that dates from this period is the Canal de Languedoc, which connects the Atlantic Ocean with Mediterranean Sea across the length of southern France.

Freight ca be carried by water much more cheaply than on land, therefore canal building has continued up to the resent time despite the attention given to railroads in the nineteenth century and to highways in the twentieth. Most of the rivers and ports of Europe are connected by a network of canals that carry a large proportion of the commerce of this highly industrialized region. In the United States, canals have less important than in Europe and China .One noteworthy nineteenth century project, however, was the Erie Canal m, which connected Albany on the Hudson River with the Great Lakes across the length of northern New York State. New York City in large part owes its commercial preeminence to the Erie Canal.

Lesson 3 The Saint Lawrence Seaway

The St. Lawrence River is arguably among the world's most unique rivers, in part because about half its discharge originates from the huge Laurentian Great Lakes and their many tributaries. This river is also known as River Saint-Laurent among the Francophones of Quebec and as Kaniatarowanenneh (approximately translated as "the big waterway") in the Iroquois language. It ranks among the top 16 rivers of the world in annual discharge and has the second-highest flow in North America (if the Saguenay is included). Hydrological contributions of the voluminous Great Lakes produce a river characterized by clearer water and a more stable stage level than any other large river in North America.

The main stem flows 965 km along a northeasterly path from the outlet of Lake Ontario to the mouth of the Saguenay River, which is the approximate downstream terminus of the upper estuary. Along this pathway the river is bounded solely by Quebec, Ontario, and New York (in order of shoreline length). Another >300 km of the St. Lawrence River system constitutes the lower estuary.

1.Physiography and climate

The shoreline along the river's nearly 1000 km main stem and the basin's contributions to water chemistry are influenced by two major physiographic divisions: the Precambrian Shield and the Appalachian Highlands, although the river as a whole is affected by three major physiographic divisions and eight physiographic provinces because it draws water from watersheds around the Great Lakes. The geology of the Precambrian Shield is dominated by silicate rocks and includes metasediments (e.g., quartzite, crystalline limestone), metavolcanics, igneous rocks (e.g., granite), orthogenesis, and pegmatite. Found in the St. Lawrence Lowlands are formations of early Paleozoic rock (Cambrian and Ordovician) lying on Precambrian bedrock, with a thick horizontal deposit of sandstone, limestone, dolomite, and shale. The remaining provinces in the Appalachian Highlands are composed of sedimentary and volcanic (igneous) rock that underwent complex metamorphic deformation during the Paleozoic era. Bedrock outcrops in the landscape surrounding the St. Lawrence River blend with flat plains, some rolling hills, and the ancient Adirondack Mountains southeast of the river.

Temperatures and precipitation in the valley of the St. Lawrence River are more stable than for rivers situated farther inland because of the influences of the Great Lakes and ocean. Moreover, precipitation is spread relatively evenly throughout the year. Montreal averages 94.2 cm of precipitation per year (snow converted to rain equivalence) and has a monthly range of 6.6 cm in February to 9.1 cm in August. The historical average temperature is 6.2 ℃ to 6.8 ℃ , with mean monthly temperature ranging from -10℃ in January to 21℃ in July. As a result of the seasonal temperature pattern, the river is at least partially ice covered for most of the winter, especially in the fluvial lake regions. In Massena, New York (near the end of the international portion of the river), the main channel is covered with ice by late December to late January, snowfall averages 178 cm/yr, and the crop-growing season is about 139 days.

2.River geomorphology and hydrology

The main stem of the St. Lawrence begins at the outlet of Lake Ontario and flows through four sections before reaching the Gulf of St. Lawrence: fluvial section, fluvial estuary, upper estuary (including the Saguenay River), and lower estuary (not discussed in this chapter). From Lake Ontario to the end of the lower estuary the river drops about 184 m, for an average of 14 cm/km. The maximum topographic relief in the basin of the St. Lawrence main stem is 1270 m, but this occurs near the downstream end of the upper estuary. Many shoreline areas of the main stem feature large boulders, but the channel is composed primarily of sand and gravel except where the natural falls formerly occurred.

The 655 km long fluvial section, which extends from Lake Ontario through the international portion of the river to Cornwall-Massena in Quebec, includes uppermost braided regions, constricted channels, rapids (now mostly bypassed with navigation locks), modest floodplain areas, and four natural fluvial lakes, often widened and deepened by hydroelectric dams. A shipping channel with a minimum depth of 8.2 m is maintained throughout this section of the river. This channel is deeper than the 3 m minimum navigation depth maintained for many other navigable rivers in the United States because oceangoing ships rather than just barges traverse the St. Lawrence Seaway. In contrast to the pool and riffle-run sequence in many rivers of North America, the natural fluvial lakes of the St. Lawrence River (i.e., Lake St. Lawrence and Lacs Saint-François, Saint-Louis, and Saint-Pierre) are relatively shallow (80% of the area<6 m) in comparison to the generally deeper, main riverine channels (often 10 to 12 m deep). These wide, shallow pools were present prior to construction of the hydroelectric and diversion dams, but the dams have modified the size and depth of these more lacustrine portions of the river.

Annual discharge of the St. Lawrence averages 12101 m^3/s at Quebec City (1962—1988) but this rises to 16800 m^3/s downstream of the confluence with the Saguenay River. The outflow of Lake Ontario averages 7410 m^3/s. The Great Lakes contribute 61% of the water reaching Quebec City and just under half the freshwater entering the Gulf of St. Lawrence. Most tributary inputs occur downstream of the international portion of the river. For example, at Massena, New York, and Cornwall, Ontario, which are very near the downstream end of the international section of the river, 99% of the discharge originates from Lake Ontario. The longest-term discharge record for the St. Lawrence is at Massena-Cornwall, where peak discharge occurs from April through June in response to snowmelt, and minimum flows are in January. However, because this fluvial section of the river is so strongly influenced by discharge from Lake Ontario, it fluctuates to a very minor degree compared to other large rivers in North America. Indeed, the average maximum daily discharge from 1900 to 1989 was 11337 m^3/s at Cornwall, less than three times the minimum daily flow of 4170 m^3/s over the same period. In a normal year, the minimum flow is only 19.5% less than the maximum flow in the fluvial section of the river, and the difference between average flow and peak flow is only 6.4%. This low variability is due mostly to the stabilizing influence of the high water volume held by the Great Lakes, where estimated water residence times for the five major lakes vary from 2.6 to 117 years. Runoff is relatively low and, like precipitation, is consistent among months (2.12 to 2.65 cm/mo) compared to many other rivers throughout North America. Farther downstream the river is more affected by seasonal flows and thus monthly variability increases. Indeed, at Quebec City, near the end of the fluvial estuary, the maximum monthly flow over the period from 1962 to 1988 was 62.2% higher than the minimum flow and 40.7% higher than the mean flow.

3.Human impacts and special foeatures

The international St. Lawrence Seaway became an important commercial link between the Great Lakes and the Atlantic Ocean, with 37 million metric tons of cargo passing the Lake Ontario-Montreal section in 1997 alone. This population density and commercial activity have come with an ecological price tag, however, both because of the construction and operation of several major hydroelectric dams, navigation locks, and diversion dams and because of pollution from industrial, municipal, and agricultural sources.

Although environmental laws now exert a stronger control over point-source and nonpoint-source pollution, regulation of the river channel continues unabated and with little significant attention to environmental consequences. Four hydroelectric dams are present on the river's main stem, and numerous channel-control structures redirect flow. Small dams on the St. Lawrence's tributaries are also quite common. In addition to these hydroelectric dams, seven navigation locks enable passage of ships through the 68 m elevational drop from Lake Ontario to the fluvial Lac Saint-Pierre, and a minimum flow channel of 8.2 m is maintained for passage of deep draft, seagoing commercial ships. The binational International Joint Commission (IJC) was established by the Boundary Waters Treaty of 1909 for the purpose of regulating aquatic-impact structures within the Great Lakes and the St. Lawrence River and for regulating the level and flow of these bound- ary waters. As a result of this regulation, the annual stage variation at the mouth of Lac Saint-François has been reduced from 60 cm to 15 cm.

Commercial traffic on the river is vital to the economy of Canada and the United States. An average of more than 10000 trips per year are made by commercial vessels, and in 1997 alone, for example, 37 million metric tons of cargo passed through the Lake Ontario-Montreal section. An increasing number of passenger ships (including pleasure ships from Europe) traverse the St. Lawrence Seaway, and recreational boating continues to grow.

The effects of regulation of the river for power generation and navigation are poorly known because of inadequate ecosystem studies and the absence of sufficient preregulation environmental data. In the case of the Beauharnois Dam near Montreal, for example, 84% of the river's flow was diverted to pass through the dam, resulting in major hydrodynamic alterations of two nearby fluvial lakes and the river's bed with unknown biotic consequences. Nonetheless, it is clear that most river dams substantially alter the nature of a lotic ecosystem, making it more lentic and interfering with the movement of fishes and other fauna.

Despite imposition of dams on the main channel and many tributaries, the St. Lawrence continues to be an unusual and good quality river. In terms of biological distinctiveness, its fauna is rated continentally outstanding, and Lac Saint-Pierre is a RAMSAR World Heritage Site. Aside from its uniqueness in terms of lake origin and overall size, the St. Lawrence is less turbid than all other larger rivers of the world. On the other hand, without the navigable access to the Atlantic Ocean provided by this river, development of large cities and a strong economy along the river and around the Great Lakes would have been severely impeded over the last 200 years.

New Words and Phrase

1.arguably *n.* 可论证的
2.stem *n.* 茎
3.terminus *n.* 终点站
4.physiography *n.* 自然地理学
5.the Precambrian Shield 前寒武纪地盾
6.the Appalachian Highland 阿巴拉契亚高地
7.watershed *n.* 流域
8.geology *n.* 地质
9.quartzite *n.* 石英岩
10.metavolcanics *n.* 变质火山岩
11.igneous *n.* 耐久性
12.landscape *n.* 陆上风景
13.geomorphology *n.* 地貌学
14.hydrology *n.* 水文
15.lacustrine *adj.* 湖泊的
16.unabated *adj.* 不减；未变弱
17.nonethless *adv.* 尽管如此
18.lentic *adj.* 静水的
19.fauna *n.* 动物群
20.imped *vi.* 冲击

Notes

1.Temperatures and precipitation in the valley of the St. Lawrence River are more stable than for rivers situated farther inland because of the influences of the Great Lakes and ocean.

受五大湖和海洋的影响，相较于内陆河流而言，圣劳伦斯河拥有更加稳定的气温和降水。

inland 内陆 valley 河谷

Arch dams transfer the greater proportion of the water load to the valley sides rather than to the floor.

2.These wide, shallow pools were present prior to construction of the hydroelectric and diversion dams, but the dams have modified the size and depth of these more lacustrine portions of the river.

相对宽浅的水潭在河流上水电和分流大坝修建前就存在，但是这些坝的修建可以改善河流部分河段的尺寸和深度。

prior to 在……之前

Prior to World War I, hydropower development remained in private hands under government regulation.

3.The effects of regulation of the river for power generation and navigation are poorly known because of inadequate ecosystem studies and the absence of sufficient pre-regulation environmental data.

水利发电站和航道修建的影响认识不足，主要是由于缺乏生态方面的研究和前期环境监测的数据。

power generation 发电 navigation 通航

In-stream use includes the use of rivers for navigation, hydroelectric power generation, fish and wildlife habitats, and recreation.

Comprehensive Exercises

I.Answer the following questions on the text.

1.What is the physiography in the St. Lawrence River?

2.What is the main characteristics of the St. Lawrence River?

3.How do the government regulate the ecological environment?

4.How do we define the criterion to estimate the quality of the St. Lawrence River?

5.What did the construction of the St. Lawrence River result in?

II.Fill the most appropriate words or phrases in the correct forms in the blanks from the list below.

Originate	contribute to	relief	navigable	in contrast to
compared to	runoff	regulation	in the case of	interfere with

1.Hard-surfaced linings are frequently protected from uplift by automatic______valves placed in the bottom and sides of the canal.

2.It is clear that most river dams substantially alter the nature of a lotic ecosystem, making it more lentic and ________the movement of fishes and other fauna.

3.Regularly flowing rivers and streams that _______within arid lands are known as "endogenous" .

4.Many different processes and materials may pollute surface water or groundwater. All segments of our society (urban, rural, industrial, agricultural, and military) _______the problem of water pollution.

5.This great system of rivers, canals and lakes makes the entire distance from the Atlantic to Duluth Minnesota, on Lake Superior, _______ for ocean-going ships.

6.Open channels are characterized by a free water surface, ______pressure conduits, which always flow full.

7.They are small _______their capacity, easy to operate, and suitable for handing sediment and other foreign material.

8.Water balance estimations are based on making river runoff maps and data for the _______of Polar glaciers more accurate.

9.______of flow in the canal and the distribution of the water are facilitated by various structures.

10.The minimum maturity on which loans will be sought is nineteen years in the case of loans guaranteed by export credit arrangements and eight to ten years______commercial bank loans.

III.Translate the following sentences into Chinese from the text.

1.The shoreline along the river's nearly 1000 km main stem and the basin's contributions to water chemistry are influenced by two major physiographic divisions: the Precambrian Shield and the Appalachian Highlands, although the river as a whole is affected by three major physiographic divisions and eight physiographic provinces because it draws water from watersheds around the Great Lakes.

2.The 655 km long fluvial section, which extends from Lake Ontario through the international portion of the river to Cornwall-Massena in Quebec, includes uppermost braided regions, constricted channels, rapids (now mostly bypassed with navigation locks), modest floodplain areas, and four natural fluvial lakes, often widened and deepened by hydroelectric dams.

3.In contrast to the pool and riffle-run sequence in many rivers of North America, the natural fluvial lakes of the St. Lawrence River (i.e., Lake St. Lawrence and Lacs Saint-François, Saint-Louis, and Saint-Pierre) are relatively shallow (80% of the area<6 m) in comparison to the generally deeper, main riverine channels (often 10 to 12 m deep).

4.This commercial activity have come with an ecological price tag, however, both because of the construction and operation of several major hydroelectric dams, navigation locks, and diversion dams and because of pollution from industrial, municipal, and agricultural sources.

Reading Material Ottawa River

The Ottawa River, or Kichesippi (meaning "Great River" in the language of the Algonquin Nation), originates in Lake Temiskaming (or Lac Temis-camingue) and flows 1271 km across the Precambrian Shield to its confluence with the St. Lawrence River near the Montreal archipelago.The Ottawa is the largest tributary in the freshwater fluvial St. Lawrence River-Great Lakes system, with a basin of 146334 km^2. It is roughly 65% in Quebec and 35% in Ontario, and the river forms the border between these two provinces for most of its length. Although much of the river flows through a naturally constricted channel, there are several complexes of islands and bays, such as the Petrie Island Preserve in the lower river. These environmentally sensitive areas provide a diverse and highly productive habitat for fish, invertebrate, and macrophyte assemblages.

Compared to southern tributaries of the St. Lawrence River, the Ottawa River is softer and lower in alkalinity but higher in nutrients and organic carbon. This reflects the primary origin of this north-ern tributary in the crystalline Precambrian Shield as well as the intensive agricultural activity in its lower basin. At its confluence with the St. Lawrence River, the more turbid waters of this northern tributary are often referred to as "brown waters" in comparison to the "green waters" derived from the Great Lakes. Despite the natural tendency of the St. Lawrence River to mix with waters of the Ottawa River, relatively distinct bodies of green and brown water can be detected at least 100 km below their confluence.

1. Physiography, climate, and land use

The basin, which is located along the southern edge of the Precambrian Shield-Frontenac axis, contains three physiographic provinces: the St. Lawrence Lowland (SL), Laurentian Highlands (LU), and Superior Upland (SU). Lowlands of the Ottawa River basin, once the floor of the ancient Champlain Sea, are bounded by two mountainous regions: the Laurentians to the west and the Algonquin dome to the east. At 968 m asl, Mount Tremblant, located just north of Montreal in the Laurentian complex, is the highest named peak in the basin and a popular skiing area. The underlying geology is dominated by a base of crystalline and crystallophyllian Precambrian rock in the north (99%) and Ordovician sedimentary rock in the south (98%). Surface soils reflect a gradation between distinctive soils in the upper and lower basin. The upper basin is characterized by well- drained organic mesisols and podzols interspersed with silt and sand deposits from the prehistoric Champlain Sea. In contrast, the lower basin features well-drained podzols interspersed with poorly drained melanic brunisols near the confluence with the St. Lawrence in the Montreal archipelago.

The climate in the Ottawa River valley is best described as humid continental. Mean annual daily temperatures are 6℃ , and the basin averages 160 to 210 growing days with temperatures of at least 5℃ . Mean monthly temperatures range from -11℃ in January to 21℃ in July. Annual precipitation averages 100.2 cm and is evenly distributed throughout the year, with rarely more than 25% falling as winter snow.Annual productivity of these second-and third-growth stands range from 200 to 500 $g \cdot m^{-2} \cdot yr^{-1}$.Most of the Ottawa River basin is forested (86%), with the remaining land cover consisting of surface waters (10%), urban areas (2%), and agriculture (2%). Most agricultural activities are clustered in the lower portion of the basin. The largest city in this watershed is Ottawa, the nation's capital, with 1.06 million.

2.River geomorphology, hydrology, and chemistry

From its origin at Lake Temiskaming, the Ottawa River flows mostly through a naturally constricted channel interrupted in some parts by wider, shallow floodplains and island mosaics. Along this pathway the main channel slopes about 36 cm/km, whereas the maximum topographic relief in the basin is 911 m. In the upper and middle sections of the river, the constricted channel flows between artificial reservoirs. The lower section of the Ottawa River is also mainly constricted but includes some floodplain regions and areas replete with islands, especially in the section extending from its confluence with the Gatineau River downstream to the confluence with the St. Lawrence River.

Geologic formations of the Precambrian Shield strongly influence the chemistry of this river. Mean conductivity (80 ms/cm), alkalinity (19.2 mg/L as $CaCO_3$), hardness (29.2 mg/L as $CaCO_3$), suspended solids (6 mg/L), turbidity (4 NTU), and dissolved organic nitrogen (0.18 mg/L) are all low compared to waters of the Great Lakes and southern tributaries of the St. Lawrence. Conversely, levels of total organic phosphorus (0.053 mg/L) and organic carbon (5.6 mg/L) are higher than those other waters. Suspended detritus concentrations increase downstream, peaking in the faster-moving middle reaches. Concentrations then drop along with current velocities as the river approaches the major hydroelectric facilities at Carillon near the confluence with the St. Lawrence. Transit time from upper tributaries to Carillon is 14 days. No consistent longitudinal patterns are evident in the mean annual levels (1979-1994) of NH_4–N (0.045 mg/L), NO_3-N+NO_2-N (0.17 mg/L), total nitrogen (0.41 mg/L), total phosphorus (0.029 mg/L), and chlorophyll a (1.9 mg/L).

3. Human impacts and special features

The Ottawa River is now generally considered to be in good condition following extensive environmental controls in the last 20 years. Prior to the early 1980s, the lower and middle reaches of the Ottawa River, with over 100 municipalities and 2000 farms in the drainage basin, had problems with elevated fecal coliform counts and nutrient inputs, especially downstream of metropolitan Ottawa-Hull and Montreal. Since then, more secondary treatment systems have come on-line and stricter regulations and installation of industrial wastewater treatment systems around the city of Gatineau have reduced inputs of copper, lead, and aluminum and cut emissions of persistent organics, such as PCBs. However, the greatest human impacts on the Ottawa River are probably its dams. Using criteria of Dynesius and Nilsson (1994), the river would be considered strongly fragmented because of the seven major dams on the main stem and over 300 impoundments on tributaries.

Lesson 4 Expansion of the Panama Canal

With the expansion of the Panama Canal, port cities along the U.S. Eastern Seaboard-from New York to Miami to Houston are competing to attract the increased trade expected once the project is complete in 2015. Of these ports, only Norfolk currently has the channel depth and landside infrastructure to receive the Post-Panamax ships (with a fifty-foot draft and a 12500-TEU[2] container capacity) that will pass through the expanded Canal. Other East-Coast ports are either engaged in, or proposing projects that include port expansion, dredging, and multimodal transportation for greater hinterland connectivity, which are estimated to cost $20 billion. Ninety percent of world trade volumes move by sea, and an American Association of Port Authorities

(AAPA) report shows that in 2011 U.S. ports generated more than 13 million direct and indirect jobs, $650 billion in personal income, and $212 billion in U.S. federal, state, and local taxes .

1.Competition and technological change in port cities

Competition among cities and regions to attract global trade flows involves development strategies that seek to offer the appropriate infrastructural hardware and capacity, either offering better hinterland market connectivity or more competitive characteristics as a transshipment base. Over the last half-century, the streamlining process of containerization allowed shipping to go beyond traditional hinterland boundaries and into new territories to compete. Container trade throughout the United States surged from the period between 1997 and 2006, going from 14.9 million to 27.4 million TEUs, based largely on strong increases in Asian imports. There are three routes for Asian cargo to reach the U.S. East Coast: the "land bridge" route from West-Coast ports across the continental U.S., the "all water" route via the Panama Canal, and the "all water" route via the Suez Canal. During the 1990s, increased Asian container trade generally berthed at West Coast ports, and often used "land bridge" train services to ship East Coast goods across the country. Double-stacked freight trains policies established in 1984 made this option more efficient and attractive for shippers. The economies of scale in using Post-Panamax vessels to transport Asian goods to U.S. markets were more cost-beneficial than sending smaller Panamax vessels through the Panama Canal to get to Eastern markets. For ports, however, larger ships meant fewer stops, and this increased inter-port competition for shipping business.

Throughout much of the twentieth century, and into the twenty-first century, a large portion of the risk that major U.S. ports face is mitigated through federal subsidies or the issue of tax-exempt debt for state and local authorities and port authorities. Ports were made public largely to eliminate private railroad monopolies of intermodal, waterfront activities, and port authorities were the semi-autonomous entities established to professionalize the sector and tie port activity to larger regional economic development objectives. In addition, it is generally agreed that ports provide facilities for international trade that benefit the economic wellbeing of the entire nation, and thus, full cost for their maintenance should not only be born by local actors. The difficulty comes in evaluating and measuring exactly how much these federal subsidies should be, given that the cost-benefit project analyses, particularly with regards to positive regional externalities and job creation, are not sophisticated enough. In this climate, shipping companies, as mobile capital, have the advantage as private actors to benefit from port competition, whereas ports essentially geographically-fixed capital seek to court shipping activity for ancillary business, job creation, and taxes by leveraging the most advanced port technology and streamlined access. Nationally, then, debates around excessive or duplicate regional funding for port expansion center on present and future capacity for trade growth. Port expansion boosterism, promoted by various stakeholders, often cultivates fear that infrastructure deficiency could slow international trade, and by extension, regional and local economic development. Opposition to specific port expansion can come from environmentalists, competing ports, and fiscal conservatives, advocating the need for more efficient, effective, strategic, and measurable public investment in ports.

2. The Panama Canal

The Panama Canal is a geotechnical, geopolitical wonder that opened in 1914—at the end of a great global economic expansion and the dawn of international political turmoil—to connect the Pacific and Atlantic Oceans. The Canal nears fifty miles in length, and is comprised of a system of artificial lakes, channels, and locks Today, more than 14000 ships pass through the Canal annually, and seventy percent of container cargo on those ships is either imported from or exported to the United States (Ibid.). The scale of super oil tankers has exceeded the Canal's dimensions since the 1960s, and Post-Panamax vessels since then, so in 2005 the decision was finally made to expand the Canal's capacity.

In 2007, Panama's vote on the issues made it official. The $5.25 billion expansion project will construct two new lock complexes on the Atlantic and Pacific entrances, deepen and widen new access channels to these locks, and deepening the navigation channel and raise the level of Gatun Lake. Presently, the Panamax vessel is the largest ship that can pass through the Canal, with a dimension of 965-feet in length, 106-feet wide, and a draft of 39.5-feet, with capacity for 4500 TEUs. The new dimensions will allow Post-Panamax vessels to pass, with dimensions of 1200-feet in length, 160-feet wide, and a draft of 50 feet, with capacity for approximately 12000 TEUs, In July 2013, Maersk launched the new 18000-TEU "Triple-E" container vessel built by the Korean Daewoo Shipbuilding and Marine Engineering, with dimensions of 1300-feet long, 193.5-feet-wide, and a draft of fifty-two-and-half feet. The Panama Canal will be unable to handle these new dimensions in its current 2015 expansion. Originally projected for 2014 completion, slowed construction pushed the expansion back to June 2015. While seventy percent of the project is complete, current financial disagreements among stakeholder due to $1.6 billion in cost overruns risk pushing the project back by up to three years to 2018.

3. Savannah and the port

The City of Savannah was founded in 1733 by General James Oglethorpe on the southern bank of the Savannah River at the Yamacraw Bluff, approximately twelve miles inland from the river mouth. Among other factors, the elevated site was selected for the protection that it provided from the hostile Spanish forces further south along the Atlantic Coast.

The Savannah River was originally maintained by the City of Savannah in collaboration with private companies, with key obstructions being shoaling, sandbars, and debris and wreckage from the Revolutionary War, the War of 1812, and the Civil War. In 1826, city aldermen sent a request to the U.S. Congress for $50000 to clean the river. The request was approved, and the money was appropriated for over thirty years to engineer the river to navigable widths of up to 300 feet, and deepen the river from 8 feet to 17 feet. The U.S. Army Corps of Engineers Savannah District was established in 1829 for the survey of the Savannah River and the construction of Fort Pulaski. Responsibility for river maintenance would eventually be passed onto the U.S. Army Corps of Engineers Savannah District. By the mid-twentieth century, the boll weevil had decimated the Southern cotton crop, and Savannah had a nascent industrial corridor along the river that included a sugar refinery and a paper mill.

The Georgia Port Authority (GPA) was created by the State in 1945, and the State purchased a 407-acre U.S Army Depot just north of Savannah, which opened as the Garden City Terminal in 1953. In 1958, the GPA purchased the Ocean Terminal from the Central Georgia Railroad, located nearer to the city under the Eugene Talmadge Bridge, and services began shortly afterward. The 1954 Rivers and Harbors Act passed by the U.S. Congress authorized the Corps to widen the Savannah River to 400 feet, and deepen it to 44 feet (Ibid.). In 1973, a 140-acre Liquid Natural Gas (LNG) facility was constructed on Elba Island, located five miles downstream of Savannah and seven miles from the Savannah River mouth. The LNG facility would slowly be expanded, and it is now one of the world's largest. Shell and Kinder Morgan, who own the plant, are applying for permits to export LNG from the Elba Island facility, which, along with the prospect of SHEP's environmental impacts, may present further environmental and safety concerns for the river channel and the wider metropolitan region. U.S. Coast Guard regulations require that other maritime traffic essentially stop when LNG is moving through a shipping channel, so the prospect of increased LNG movement through the Savannah shipping channel would increase choke-point delays for container throughput; the same choke-points that are cited as one of the principal arguments in favor of the SHEP dredging. Indeed, competing infrastructure projects at the local level in Savannah risk the kind uncoordinated national, coastal, and regional competition explored throughout this paper.

Today, Metropolitan Savannah has a population of just over 350000. The Port of Savannah has been the fastest- growing container port in the country since 2001, and in 2006 it overtook Charleston, S.C. to become the top container port of the Southeast, moving 2982467 TEU containers in FY 2012). Charleston's port has plans to dredge its harbor to 50 feet, in hopes of reclaiming the top position. Savannah's Garden City Terminal benefits from on-site intermodal Class 1 rail service from both CSX Transportation and Norfolk Southern Railroad. For inter-modal trucking, the terminal has easy access to the principal North-South coastal Interstate 95 (5.5 miles), and Interstate 16 (5.7 miles), with direct access to Atlanta. The Garden City Terminal is the largest-single terminal facility in the U.S., with 1200 acres, and nine berths totaling 9693 feet. Savannah emphasizes a "business-friendly" environment (Georgia is a "right to work" state), and the Savannah Economic Development Authority developed the Crossroads Business Center-a 1700-acre logistics park in the late 1990s to help attract businesses. This initial project helped to spark regional distribution center development, which includes companies such as Home Depot, Lowes, Tar-get, Pier 1 Imports, Coca-Cola, and Ikea. Savannah now boasts the largest concentration of retail import distributions centers on the East Coast, and Georgia leads the nation in warehouse and distribution center development, promoting construction of a total of 3.4 million square-feet of new warehousing throughout the state. The Port of Brunswick, located eighty miles south of Savannah along the Georgia coast a, is the third busiest U.S. port for auto imports. The GPA is owner and operator of both the Savannah and Brunswick ports.

Post-Panamax ships coming through the Suez Canal are currently berthing at the Savannah port with eight trade route services, but to do so they must incur additional operating costs by either lightening their maximum loads before entering the channel, or waiting for the limited tidal window that permits deeper-berth vessels . Georgia's strong investment in trade infrastructure is a key part of its economic development strategy.

New Words and Phrase

1.engage *vi.* 聘用，与……建立关系
2.appropriate *adj.* 合适的，恰当的
3.multimodal *n.* 多峰
4.containerization *n.* 集装箱化
5.hinterland *n.* 内陆；腹地
6.waterfront *n.* 滨水区，码头区
7.professionalize *vi.* 使专业化，使职业化
8.leverage *n.* 影响力，杠杆作用效力
9.boosterism *n.* 支持主义
10.fiscal *adj.* 财政的，国库的
11.hostile *adj.* 敌对的，坚决否定
12.collaboration *n.* 合作，协作
13.reclaim *vi.* 取回，拿回
14.vessel *n.* 大船，轮船
15.promote *vi.* 促进

Notes

1.With the expansion of the Panama Canal, port cities along the U.S. Eastern Seaboard-from New York to Miami to Houston are competing to attract the increased trade expected once the project is complete in 2015.

随着巴拿马运河扩能工程的开展，美国东海岸沿线从纽约至迈阿密再到休斯顿港口将开始竞争从而吸引 2015 年该工程完工后所带来的贸易增长。

once 一旦 with 随着，常接状语

Arch dams transfer the greater proportion of the water load to the valley sides rather than to the floor.

2.Competition among cities and regions to attract global trade flows involves development strategies that seek to offer the appropriate infrastructural hardware and capacity, either offering better hinterland market connectivity or more competitive characteristics as a transshipment base.

城市与区域之间的全球贸易竞争涉及追求恰当基础设施和容积，要么提供更好的与内陆市场的连通，要么作为更具竞争特色的转运基地的发展方针。

seek to 追求，追逐 either… or… 要么，要么

Prior to World War I, hydropower development remained in private hands under government regulation.

3.The scale of super oil tankers has exceeded the Canal's dimensions since the 1960s, and Post-Panamax vessels since then, so in 2005 the decision was finally made to expand the Canal's capacity.

巨型油轮的尺度从 1960 年开始超过了把运河的尺度，因此在 2005 年，超巴拿马轮船使得该运河的容量进行了扩充。

be made to power generation 被迫做某事，陈述一个事实

Includes the use of rivers for navigation, hydroelectric power generation, fish and wildlife habitats, and recreation.

Comprehensive Exercises

I.Answer the following questions on the text.

1.What do the competition among cities and regions to attract global trade involve?

2.What is the Panama Canal consist of?

3.When did the Georgia Port become the top container port of the Southeast?

4.What is the main characteristics of the Panama Canal?

5.What is the key to influence the development of the canal capacity?

II.Fill the most appropriate words or phrases in the correct forms in the blanks from the list below.

propose	seek to	authority	with regard to	promote
comprise of	handle	vessel	investment	strategy

1.In later years the _____of the Bonneville Power Administration to market power expanded to include 32 additional federal projects.

2.In the event of fire, the general_____for tunnels up to about 15 km long is to get the train out of the tunnel in order to reduce the consequences of the fire.

3. _____the construction of the casing around the impeller, centrifugal pumps are classified as volute or turbine.

4.The appropriate factor to convert _____into an equivalent annual cost is designated as the capital recovery factor and may be computed from the expression where represents the interest rate per annum and represents the years of estimated life.

5.Although lower air pressure helps______evaporation, temperature is the primary factor.

6.This paper______a new method to investigate the sediment behavior in the mountain river.

7.More and more businesses want this energy to rub off on them, and _____ generate some fireworks of their own as they discover their inner tweeters.

8.The Three Gorges Dam _____three parts: the dam itself, the hydropower stations and the navigation facility.

9.Location measured by the GPS technique is expressed in geography coordinates_____by a reference system named WGS84, which was specially created for the purpose of the GPS use.

10.In July 2013, Maersk launched the new 18000-TEU container_____built by the Korean Daewoo Shipbuilding and Marine Engineering, with dimensions of 1300-feet long, 193.5-feet-wide, and a draft of fifty-two-and-half feet.

III.Translate the following sentences into Chinese from the text.

1.There are three routes for Asian cargo to reach the U.S. East Coast: the "land bridge" route from West-Coast ports across the continental U.S., the "all water" route via the Panama Canal, and the "all water" route via the Suez Canal.

2.Ports were made public largely to eliminate private railroad monopolies of intermodal, waterfront activities, and port authorities were the semi-autonomous entities established to professionalize the sector and tie port activity to larger regional economic development objectives.

3.Coast Guard regulations require that other maritime traffic essentially stop when LNG is moving through a shipping channel, so the prospect of increased LNG movement through the Savannah shipping channel would increase choke-point delays for container throughput; the same choke-points that are cited as one of the principal arguments in favor of the SHEP dredging.

4.Post-Panamax ships coming through the Suez Canal are currently berthing at the Savannah port with eight trade route services, but to do so they must incur additional operating costs by either lightening their maximum loads before entering the channel, or waiting for the limited tidal window that permits deeper-berth vessels.

科技英语名词化结构

科技英语的两个最显著特点：广泛使用名词化结构和大量使用被动语态。名词化结构指的是大量使用名词和名词词组即在日常英语或其他功能和题材里用动词、形容词等词类充当某种语法成分，在科技英语里往往会转化为由名词充当这种语法成分。这种名词化结构可以用以下一组例子来说明。在日常英语中我们通常说：

(1)You can rectify this fault if you insert a slash.

插入一条斜线便可以纠正此错误。在科技英语中则通常说：

(2)Rectification of this fault is achieved by insertion of a wedge.

在这组例子中，名词化发生在两个地方。首先，例 (1) 中的谓语动词转化为 (2) 中的名词主语，同时增加了一个内容空泛的动词来充当谓语动词；其次，例 (1) 中的状语从句的谓语动词 insert 转化为名词结构介词短语来充当状语。经过名词化处理的例 (2) 由含有两个主谓结构的复合句变成了只含一个主谓结构的简单句，从而使句子的结构更加精练严谨，也由于使用抽象名词替代原来的人称代词做主语而使句子的语体更加正式。下面一组例子属于同样情况：

(3)We can improve its performance when we use superheated steam.

可以使用超热蒸气改进其性能。

(4)An improvement of its performance can be effected by the use of superheated steam.

动词名词化的结果是科技英语中行为名词的出现频率大大增加。这些行为名词可以由动词加上 _ment，_sion，_xion，_ment，_ance，_ence 等后缀构成。这类名词除表示行为动作外，还可表示状态、手段、结果及存在，其中以 _ment 为后缀的词还能表示事物或工具。例如：

(5)The dependence of the rate of evaporation of a liquid on temperature is enormous. 液体蒸发速度很大程度上取决于它的温度。

(6)The construction of such satellites has now been realized，its realization being supported with all the achievements of modern science. 由于得到现代科学所有成就的支持，现在制造了这样的卫星。

(7)A number of energy conversion systems are now in transportation . 许多能量转换系统现在已用于运输方面。

在某些情况下，形容词也可名词化。如：

(8)It is doubtful how accurate the results are. 结果的正确性值得怀疑。也可表示为（9）。

(9)The accuracy of these results is doubtful.

有时从句也可以名词化。下面例句中的 If 从句能变成担任主语的名词词组。

(10)If we add or remove heat，the state of matter may change . 添加或者减少热量可以改变物体的状态。

(11)The addition or remove of heat may change the state of matter.

表示行为或状态的名词一般属于抽象名词，科技行文的抽象程度要比表达类似内容的普通英语更高。它的大量使用表明科技文献借助于抽象思维的逻辑性和概念化，追求表述的简练、凝重、客观和浓缩，这与具体的形象描绘为主要表现手段的文学语言形成了鲜明的对比。

1. 科技英语中名词化结构的原因

科技英语文献是以事实为基础记述客观事物的。在遣词造句中要求见物，而名词正是表物的词汇，故在普通英语中用动词等表示的内容，科技英语却惯用名词来表达，而把原来的施动含义蕴藏在结构的深层里。

科技英语用词简洁、表达确切，结构严密、描述客观。名词化结构是以短语形式来表达一个句子，结构言简意赅，内部组织严密，而且可以把更多的信息结构融于一体，使彼此的逻辑关系更明确，表达得更细密。在对事物进行描述时，名词化结构是除被动语态之外用以提高客观程度不可忽视的途径。

应该指出，名词化结构较多地使用抽象名词表达动作和状态，这也是英语表达和汉语表达上的主要差别之一。在汉语中用行为动词表达的意思在英语中有时要转为名词，这种例子并不是个别的。例如，在汉语"我们学院已开始收订下一年度的杂志"这一句子里，"收订"用作动词，但由于"收订"的真正主语在句子中未出现，因此将"收订"机械地译成谓语动词会造成Chinglish 的句子，在逻辑上是不通的。一种较妥善的处理是将"收订"译成名词：

(12)Subscription for next year's magazine has begun in our college.

鉴于英语、汉语表达上的差别，在翻译科技英语时就要注意将某些抽象名词译成汉语中的动词或其他词类。

(13)Television is the transmission and reception of images of moving objects by radio waves. 电视是通过电波发送和接收活动物体的图像的装置。

一些行为名词词组如上所述，实际上是浓缩的句子，翻译时译成汉语句子，有时还要用泛指的人称添加主语，这样处理可以避免硬搬英文原文的结构而造成中文译文的累赘别扭。

2. 名词化结构的构成

名词化结构用词简洁、结构紧凑、表意具体、表达客观，而且整个句子的结构便于写作修辞，词句负载信息的容量得到了增加，有利于达到交际的目的。其经常使用的名词化结构有：

1) 名词 /(行为名词)+ 介词 + 名词

在此结构中，若"介词 + 名词"构成的介词短语在逻辑上是行为名词的动作对象或动作的发出者，行为名词的含义在深层中转换 (shift) 或变异 (variation), 使原来的名词变为动词，构成了动宾或主谓的关系，则在翻译时可以译成汉语的动宾结构或主谓结构。

(14)The flow of electrons is from the negative zinc plate to the positive copper plate.

电子从负的锌板流向正的铜板。

此句的 the flow of the electrons … =the electrons flows …

(15)Again in the case of all motor vehicles, friction is essential in the operation of the brake.

还有，所有机动车辆的制动器工作时都需要摩擦。

The operation of brake =the brake operates

(16)They are the employers of managers, as much as they are the employers of workpeople.

他们不仅雇用工人，也雇用经理。

此句等于 They employ managers as well as workpeople.

又如 :

(17)Television is the transm ission and re-ception of moving objects image by radio waves.

电视通过无线电波发射和接收活动物体的图像。

(18)Farm tractors are big users of diesel power.

农用拖拉机也大多以柴油机为动力。

在这一句话中 , 从表层结构来看 , big user 与 of diesel power 是从属关系 , 可其深层结构却是动宾补关系 , big users of diesel power = mainly use diesel as power , 所以不能译成 " 农用拖拉机也是柴油机的大用户 " 。

2) 介词 + 名词 (行为名词)

在此结构中 , 往往因行为名词的动作意义相对完整 , 与它同句中的其他部分之间存在着一定的逻辑关系 , 能起到时间状语、原因状语、条件状语和让步状语等作用 , 因此可以用介词短语来代替各种状语从句。

(19)Before germination, the seed is watered.

在发芽前给种子浇水。

此句可以改写成：" Before the plant germinates , it is watered ."

(20)A soluble crystalline solid may be separated from a solution by evaporation.

可溶性晶体可以通过蒸发从溶液中分离出来。

3) 中性名词 + 行为名词 (+ 介词短语)

此结构可以将宾语 (介词宾语) 转换成谓语。

(21)Rockets have found application for the exploration of the universe.

火箭已经用来探索宇宙。

此句 Rockets have found application for the exploration of the universe. 等于 People have applied rockets to explore the universe.

此结构中谓语动词含义空泛 , 在句子中只起语法作用 , 翻译时可以不译。类似的动词有 : do , keep , have , make , take , pay , show , perform 等。又如 :

(22)Curved rails offer resistance to the movement of the train.

弯曲的钢轨阻碍火车运行。

句中 offer 是中性名词 , 几乎不表示什么意义 , 只起连接作用。此句可以改为：Curved rails resist the movement of the train.

4) 与动词构成固定搭配

名词化结构与动词构成固定搭配的常用形式为 : 动词名词化结构 / 动词 + 介词名词化

结构。这种搭配大量地以一动词短语的形式出现，约定俗成。例如：

call attention to ...　注意……　　draw a distinction between...　区分……
lay emphasis on ...　强调……　　take possesion of ...　拥有……

5) 行为名词 + 短语 / 从句

在此结构中行为名词可以译成动词，与后面的成分一起构成汉语的动宾结构。

(23)I have a doubt whether the news is true.

我怀疑这消息是否确实。

此句可以改换成 I doubt whether the news is true or not.

6) 名词 + 名词 (行为名词)

在此结构中，名词在表层结构上是前置定语，但在翻译过程中，其深层结构的内在含义可以译成动宾词组，行为名词转换成谓语。

power generation　发电
hail prevention　防冰雹

UNIT IV MEASURMENTS IN HYDRAULIC WORK

Lesson 1 Physical River Model

Physical models of rivers have existed at least since 1875, when Louis Jerome Fargue built a model of the Garonne River at Bordeaux. Physical models are usually built to test various river engineering structures and to carry out experiments under controlled laboratory conditions as opposed to costly field programs. The main purposes of physical models include: (1) a small-scale laboratory duplication of a flow phenomenon observed in a river; (2) the examination of the performance of various hydraulic structures or alternative countermeasures to be considered in the final design; and (3) investigation of the model performance under different hydraulic and sediment conditions.

1. Hydraulic similitude

The prototype conditions, denoted by the subscript p, refer to the full-scale field conditions for which a hydraulic model, subscript m, is to be built in the laboratory. Model scales, subscript r, refer to the ratio of prototype to model conditions. For instance, the gravitational acceleration in the prototype is g_p, the gravitational acceleration in the model is gm, and the scale ratio for gravitational acceleration is defined as $g_r=g_p/g_m$. Hydraulic models usually have the same gravitational acceleration in the model and the prototype; thus $g_r = 1$.

For all scale models, the following considerations are relevant:

(1)the model length must be large enough to ensure the accuracy of the measurements, e.g., a flow-depth measurement error of 1 mm in a model at a scale of 1:100, or $z_r = 100$, represents an error of 10 cm in the prototype flow-depth measurement.

(2)we must consider the physical limitations on space, water discharge, and instrumentation accuracy, e.g., we cannot realistically model the Mississippi River in a 100 m-long hydraulics laboratory.

(3)we must appropriately simulate the boundary conditions, e.g., the stage and the discharge of inflow tributaries, and the possible tidal effects at the downstream end must be properly accounted for.

Hydraulic models use water and require that the scales of mass density ρ_r and kinematic viscosity ν_r be unity. Because the scale ratio for gravitational acceleration $g_r = 1$, the scale for specific weight γ_r and dynamic viscosity μ_r are also unity; thus $\rho_r = g_r = \nu_r = \gamma_r = \mu_r = 1$ in hydraulic models.

Geometric similitude describes the relative size of two Cartesian systems of coordinates (x, y, z). The vertical z_r length scale is defined as the ratio of the prototype vertical length z_p to the model vertical length z_m such that $z_r = z_p/z_m$. For instance, a length scale $z_r = 100$ indicates that a model length of 1 m corresponds to a prototype length of 100 m. The horizontal length scales are defined in a corresponding manner in the downstream x and lateral y directions as x_r and y_r. Exact geometric similitude is obtained when the vertical and the horizontal length scales are identical, i.e., $L_r = x_r = y_r = z_r$. The corresponding area and volume scales are respectively $A_r = L^2$ and $\mathrm{Vol}_r = L^3$. If accurate quantitative data are to be obtained from a model study, there must be exact geometric similitude in every linear dimension. Model distortion implies that the vertical z_r and the lateral y_r scales are not identical. The distortion factor is obtained from the ratio of y_r /z_r. Model tilting results from different vertical z_r and downstream horizontal x_r scales. The downstream model slope $S_r = z_r /x_r$ is effectively tilted when the horizontal length scale is different from the vertical scale. In a distorted model, the surface area scales for horizontal and cross-sectional surfaces are respectively $A_r = x_r \cdot y_r$ and $A_{xr} = z_r \cdot y_r$. The volume scale for a distorted model corresponds to $\mathrm{Vol}_r = x_r \cdot y_r \cdot z_r$.

Kinematic similitude refers to parameters involving length and time, e.g., velocity V, acceleration a, kinematic viscosity ν , etc. For instance, the velocity scale V_r is defined as the ratio of prototype to model velocities as $V_r = V_p / V_m$. The time scale $t_r = t_p/t_m$ appropriately describes kinematic similitude when fluid motion in the model and the prototype are similar. With the kinematic relationships that $L = V_t$ and $V = a_t$, time can be canceled from these two relationships to obtain $V_2 = a_L$. Any experiment in which the gravitational acceleration is the same in the model and the prototype requires that $a_r = g_r = 1$. When applied to the model and the prototype, this relationship yields one of the most important kinematic relationships in physical modeling:

This is known as the Froude similitude criterion. Accordingly, the time scale and the velocity scales for exact kinematic similitude are identical, $t_r = V_r = z^{0.5.}$ It is important to consider that the time scale for distorted and tilted models varies with direction. Distorted and tilted models are restricted to simulate 1D flows and time scales for flow velocities in the y and the z directions are irrelevant. Distorted and tilted models therefore cannot appropriately account for 2D and 3D convective and turbulent accelerations and should not be used to model vorticity, diffusion, turbulent mixing, and dispersion.

Dynamic similitude implies a similarity in the dynamic behavior of fluids. Dynamic similitude refers to parameters involving mass, e.g., mass density ρ, specific weight γ, and dynamic viscosity μ. For instance, the mass-density scale ρ_r is defined as the ratio of prototype to model mass densities as $\rho_r = \rho_p/\rho_m$. The mass scale $M_r = M_p/M_m$ appropriately describes dynamic similitude, besides the readily defined length and time scales. The basic concept of dynamic similitude is that individual forces acting on corresponding fluid elements must have the same force ratio in both systems. Individual forces acting on a fluid element may be due either to a body force such as weight in a gravitational field, or surface forces resulting from pressure gradients, viscous shear, or surface tension. The resulting inertial force necessitates that the force polygon be geometrically similar.

Gravitational and viscous effects are respectively described by the Froude number $F_{rr} = V_r / (g_r z_r)^{0.5}$ and the Reynolds number $R_{er} = V_r z_r / \nu_r$. In hydraulic models, the Froude and the Reynolds numbers, i.e., $F_{rr}=R_{er}=1$, can be simultaneously satisfied only when $V_2/z_r = V_r z_r$, which is the trivial full scale $V_r = z_r = 1$. We thus conclude that exact similitude of all force ratios in hydraulic models is strictly impossible except at full scale. Of course, forces that are negligible compared with others will not affect the force polygon. Therefore the art of hydraulic modeling is to center the analysis around the force components that are dominant in the system.

The art of hydraulic modeling thus consists of determining whether gravity or viscosity is the predominant physical parameter and to determine the scale parameters accordingly. This approach is reasonable as long as either gravitational or viscous terms can be neglected. In open channels, gravitational effects are typically predominant and resistance to flow does not depend on viscosity as long as flows are hydraulically rough. In most river models, however, the weight force ratio $F_g r = \rho_r L^3$ gr should balance the inertial or hydro-dynamic force $F_{ir} = \rho_r L^2 V^2$. The ratio of inertial to weight forces implies that $F_{ir} / F_{gr} = V^2 / L_r g_r = 1$, which is simply the Froude similitude criterion.

The Froude similitude criterion thus satisfies similarity in the ratio of inertial to weight forces. We also recognize that the Froude number also properly scales the ratio of velocity head to flow depth. Consequently Froude similitude describes similarity conditions in specific energy diagrams for the model and the prototype. The Froude similitude is therefore useful in describing rapidly varied flow conditions.

2.Rigid-bed model

Rigid-bed models are built to simulate flow around river improvement works and hydraulic structures. A rigid boundary implies that the bed is fixed, i.e., no sediment transport. This is the case when the Shields parameter of the bed material is $\tau_* < 0.03$. Rigid-bed model scales can be determined in either one of two cases: (1) exact geometric similitude, in which resistance to flow can be neglected; and (2) distorted/tilted models, in which resistance to flow is important. Exact geometric similitude and Froude similitude can be simultaneously maintained in rigid-bed models only when resistance to flow can be neglected. Such models are well suited to the analysis of 3D flow around hydraulic structures, in which sediment transport is not important. When long river reaches are considered and resistance to flow cannot be neglected, both the Froude and the resistance similitude can be simultaneously satisfied in tilted/distorted models.

Model distortion and tilting is viewed as a feasible practical alternative. Model distortion and tilting are acceptable only when vertical and lateral accelerations of the water can be neglected with respect to the gravitational acceleration. This practical solution allows the use of different scales for flow depth and sediment size. The model is distorted when $y_r \neq z_r$ and tilted when $x_r \neq z_r$, which should be appropriate for near 1D flow conditions. Rigid-bed hydraulic models require that resistance to flow be the same for the model and the prototype. In the hydraulically rough regime, $R_{e*}> 70$, resistance to flow depends on relative submergence h/d_s. The governing equation to be preserved in gradually varied flow models with rigid boundaries is the resistance relationship $S_r = f_r F_r^2$ whereby tilting is required because $F_{rr} = 1$ and $d_{sr} \neq 1$. In general terms, resistance to flow can be defined as $= a\,(h/d_s)\,m$ where $m = [1/(\ln 12.2h/d_s)]$.

As a particular case, according to Strickler's relationship between Manning coefficient n and bed roughness diameter $n \sim d^{1/6}$, the Manning-Strickler equation corresponds to $m = 1/6$.

The model scales in distorted Froude models must simultaneously satisfy the Froude and the Manning-Strickler similitude criteria. The Manning-Strickler similitude criterion in a distorted model is defined as$(Z_r/d_{sr})^{1/6}[(Z_r^{1/2}\ S_r^{1/2})/V_r\ ] =1$. A tilted hydraulic model $S_r = z_r/x_r \neq 1$ that satisfies the Froude similitude $F_{rr} = 1$ implies that $d_{sr} = z_r^{\ 4}/x_r^{\ 3}$. According to this relationship, the user has 2 degrees of freedom in selecting two of the three scale parameters, x_r, z_r, or ds_r. During the calibration of rigid-boundary models, model roughness is typically increased when disproportionately large blocks and sticks are used to reproduce a stage-discharge relationship comparable with that of the prototype. The modeling of design structures with distorted rigid-boundary models thus requires the intuition and judgement of experienced engineers.

Model distortion is often encountered in engineering practice whereby the flow depth is increased compared with that of exact similitude. A distorted model with different horizontal and vertical scales allows different scales for the bed material and for flow depth. The practical interest in distorted models is that in increasing flow depth and decreasing resistance to flow, and the model user can empirically increase the size of roughness elements until the model results compare with field measurements. The model is then said to be calibrated. Because the kinematic similarity is not exact, however, any attempt to determine kinematic properties such as streamlines and turbulent mixing cannot be properly scaled in distorted models.

New Words and Phrase

1.duplication *n.* 重复
2.alternative *adj.* 可替代的，另类的
3.countermeasure *n.* 对策
4.instrumentation *n.* 仪器
5.realistical *adj.* 现实的
6.tributary *n.* 支流
7.account for 解释，说明
8.scale *n.* 尺度
9.distortion *n.* 变形，失真
10.tilte *vi.* （使）倾斜
11.criterion *n.* 标准
12.act on 对……起作用，对……有功效
13.polygon *n.* 多边形，多角形
14.trivial *adj.* 不重要的，琐碎的
15.rigid *adj.* 一成不变的，坚硬的
16.intuition *n.* 直觉融合
17.stage-discharge relationship 水位—流量关系
18.calibrate *vi.* 验证

Notes

1.Physical models are usually built to test various river engineering structures and to carry out experiments under controlled laboratory conditions as opposed to costly field programs.

相对于高成本的现场测量而言，物理模型常用于测试各类型河流工程结构并在可控制的试验条件下开展试验。

carry out　实施；执行；实行

The experiments was carried out to investigate the sediment behavior in the Yangtze River.

as opposed to　与……截然相反；对照；与……相对；不是

2.The basic concept of dynamic similitude is that individual forces acting on corresponding fluid elements must have the same force ratio in both systems.

动力相似的基本概念是指作用于流体单元的外力应该在两个系统中具有相同的力比。

dynamic similitude　动力相似　　force ratio　力比

3.A distorted model with different horizontal and vertical scales allows different scales for the bed material and for flow depth.

横向与垂向尺度不同的变形模型允许其床沙与水深尺度亦不相同。

句中的 for flow depth 中的 for 可省略。

Comprehensive Exercises

I.Answer the following questions on the text.

1.What are the main purpose of the physical models?

2.What do we consider in the full-scale models?

3.What is the art of hydraulic modeling?

4.How to satisfy the similitude criteria in distorted models?

5.How many types of hydraulic similitude we should consider?

II.Fill the most appropriate words or phrases in the correct forms in the blanks from the list below.

carry out	investigation	relevant	account for	result from
cancel	act on	scale	inertial	compared with

1.A group of machines that are used together to _____the different steps in laying a concrete wearing surface for a highway.

2.______models of the dam are often made so that they can be tested under simulated conditions.

3._______revealed that the buckling of one of the extremely large steel compression members caused the collapse.

4.In the Warren truss, the diagonals_____ are in tension and compression.

5.The horizontal components of the thrust of the arch _____, and the foundation only has to resist the vertical components of the thrust.

6.The force of gravity, the natural pull of the earth, for example, is one of the stresses that ______an object.

7.The branch of engineering that deals with the effects and processes that _____the behavior of tiny particles of matter called electrons.

8.Before liberation, China's industry _____only about 30% of the total value of the country's industrial and agricultural output.

9._____operational data of the plant should be collected and made readily available for making operating decisions, and stored in a database for later evaluation of plant performance.

10.In undisturbed ______motion, the angular momentum is constant.

III.Translate the following sentences into Chinese from the text.

1.The prototype conditions, denoted by the subscript p, refer to the full-scale field conditions for which a hydraulic model, subscript m, is to be built in the laboratory. Model scales, subscript r, refer to the ratio of prototype to model conditions. For instance, the gravitational acceleration in the prototype is g_p, the gravitational acceleration in the model is g_m, and the scale ratio for gravitational acceleration is defined as $g_r = g_p/g_m$.

2.This is the case when the Shields parameter of the bed material is $\tau_* < 0.03$. Rigid-bed model scales can be determined in either one of two cases: (1) exact geometric similitude, in which resistance to flow can be neglected; and (2) distorted/tilted models, in which resistance to flow is important.

3.A distorted model with different horizontal and vertical scales allows different scales for the bed material and for flow depth. The practical interest in distorted models is that in increasing flow depth and decreasing resistance to flow, and the model user can empirically increase the size of roughness elements until the model results compare with field measurements.

4.It is important to acknowledge that complete mobile-bed similitude implies that the downstream direction is dominant and the accelerations in the lateral and the vertical directions are negligible. Complete mobile-bed similitude is therefore essentially suitable for 1D sediment-transport processes.

Reading Material Mobile-bed Model

Mobile-bed models are useful when sediment transport is significant, e.g., when $\tau_* > 0.06$. Typical examples include drop structures, local scour, erosion below spillways, sills, locks and dams, reservoir sedimentation, etc. The bed mobility provides 1 additional degree of freedom in selecting the mass density of sediment. Similitude in sediment transport is obtained when the Shields parameter τ_* and the dimensionless grain diameter d_* are similar in both systems, i.e., $\tau_{*r}=1$ and $d_{*r}=1$.Of course, these conditions also imply that $R_{e*Y}=1$ because $\tau_* d^3_*=R_{e*}^2$.

There are four similitude criteria for mobile-bed models: (1) Froude similitude; (2) resistance, e.g., Manning-Strickler; (3) dimensionless grain diameter; and (4) bed-material entrianment or Shields parameter. These four similitude criteria must be simultaneously satisfied in river reaches with rapidly varied flow and sediment transport. Hydraulic models with $g_r=v_r$ and four equations of similitude leave only 1 degree of freedom, e.g., the model length scale z_r, besides the lateral scale y_r, which is not specified by the equations.

The mobile-bed similitude is said to be complete, with 1 degree of freedom, when the four equations of similitude are simultaneously satisfied. When complete similitude is impossible, it is sometimes possible to sacrifice one of the governing equations for an additional degree of freedom.

1.Complete mobile-bed similitude

It is important to acknowledge that complete mobile-bed similitude implies that the downstream direction is dominant and the accelerations in the lateral and the vertical directions are negligible. Complete mobile-bed similitude is therefore essentially suitable for 1D sediment-transport processes. Similitude in dimensionless particle diameter $d_{*r}=1$ in hydraulic models imposes the following relationship between the particle diameter and the particle density:

$$d_{sr}^3=\frac{1}{G-1} \tag{4.1}$$

It is clear from this relationship that hydraulic models require very light sediment when large particles are used in the model.It is interesting to note that, for prototype sediment at a specific gravity of 2.65, any lightweight material corresponds to a specific scale ratio for the particle diameter.

Similitude in Shields parameter $\tau_{*r}=1$ imposes the following relationship between the particle diameter and the slope similitude S_r:

$$\tau_{*r}=\frac{z_r S_r}{(G-1)_r d_{sr}}=1 \tag{4.2}$$

The criterion for sediment suspension based on settling velocity ω can be defined from $\omega=8(v/d_s)\left[(1+0.0139d_*^3)^{0.5}-1\right]$. The settling velocity scale in water is thus $\omega_r=1/d_{sr}$ as long as $d_{*r}=1$. The criterion for sediment suspension is defined from identical values of the ratio of shear velocity to settling velocity, or $\omega_r/u_{*r}=1$. This leads directly to $d_{sr}=x_r^{1/2}/z_r$, which is identical to the condition previously obtained from the Shields parameter. We can thus conclude that similitude in Shields parameter is equivalent to similitude in the ratio of bedload to sediment suspension, provided that $d_{*r}=1$. This strengthens the requirement that $d_{*r}=1$ and $\tau_{*r}=1$ for similitude in sediment transport. Similitude in bedload sediment transport can be determined from the Einstein-Brown relationship as $q_{bv}/\omega d_s = f(\tau_*)$, in which, with $\omega_r=1/d_{sr}$ when $d_{*r}=1$, we obtain directly $q_{bvr}=f(\tau_{*r})$. It is interesting to note that $q_{bvr}=1$ when $\tau_{*r}=1$, which again strengthens the requirement for similitude in Shields parameter through $\tau_{*r}=1$.

Bed aggradation and degradation relates to the sediment continuity relation- ship applied to bedload discharge written in 1D form as $\partial q_b/\partial x=-p_0(\partial z_0/\partial t_s)$, where q_b is the unit bedload discharge, p_0 is the porosity of the bed material, z_0 is the bed-elevation, and t_s refers to time. The time scale for bedload motion t_{sr} that describes bed-elevation changes is then obtained as $t_{sr}=[(p_{0r}z_r x_r)/q_{br}]$. It can be assumed that the porosity ratio $p_{or}=1$. This time ratio refers to the sedimentation time scale that is useful in the analysis of local bed-elevation changes through local scour, bedforms, and changes in bedload transport. The time scale for bed-elevation changes is different from the time scale obtained from the Froude similitude criterion.

In diffusion-dispersion studies, the time scale for vertical mixing can be estimated from $t_{vr}=z_r/u_{*r}$ comparatively with the time to lateral mixing given by $t_{tr}=y_r^2/z_r u_{*r}$. Of course these two scales are equivalent only as long as the model is not distorted. Also, the length scale for vertical mixing is $x_{vr}=[(z_r V_r)/u_{*r}]$ is comparable with the length for lateral mixing given by $x_{tr}=V_r y_r^2/z_r u_{*r}$. These length scales are compatible for only undistorted models.

2.Incomplete mobile-bed similitude

When the conditions for complete similitude are not practically possible, one constraint can sometimes be sacrificed in order to benefit from an additional degree of freedom. As the model further deviates from complete similitude, there is a greater risk that the model may yield incorrect results. There are nevertheless a number of possibilities, depending on which conditions of similitude should be preserved in both the model and in the prototype. Two types of models are considered here: (1) non-Froudian similitude $Fr_r \neq 1$; and (2) quasi similitude in sediment transport $d_{*r} \neq 1$. First, near-equilibrium streams in which the flow is gradually varied can be simulated with different values of the Froude number as long as the flow is subcritical, i.e., fine-grained alluvial rivers with low Froude numbers can be simulated with the same bed material at higher, yet subcritical, Froude numbers. Second, coarse bed material in which bedload transport is predominant can be simulated with smaller values of d_* as long as the flow is hydraulically rough.

Lesson 2 Field Measurements

Modern theoretical assessment of bed load was initiated by Du Boys in 1879 but according to Bogardi the first attempt to measure bed load volume was carried out in the River Mur in Styria, 29 years later. In geology and geomorphology, the only "complete" measurement of bed load volume, including particle properties, is applied in river deltas entering lakes or reservoirs. There is a continually growing list of important publications dedicated to bed load functions, but data for validation is spare and mostly restricted to flume experiments. This situation is further complicated by divisions between engineers and geoscientists. New approaches are not only essential for improving models, they are also essential for improving bed load measuring techniques.

1. Traps and samplers

Since the beginning of construction of hydropower schemes, the demand for improved information on bed material transport has increased. A bag was mounted onto a frame, or sediment was trapped in mesh-covered boxes suspended from bridges. For finer particles, such as sand and gravel, these approaches were replaced by pressure difference samplers. The entrance of these samplers was designed to allow the water velocity to remain unaltered relative to the ambient velocity. In Europe, this sampler type is known as the VUV sampler and was further developed into a standard instrument by the US Geological Survey. For the latter, some calibration was even carried out under field conditions. Today these samplers are commonly applied by river authorities all over the globe. The main restriction of this approach is the nature of unsteady bed load movement. Particles frequently move along preferred lines or bed load streets and appreciable amounts of data collected by lengthy measurement programmers are required to develop a reliable picture of the natural situation.

2.Acoustic detectors

Acoustic detectors or hydro-microphones measure the noise of inter particle collisions as an indirect result of bed load transport. There has been a long list of successors attempting to define relations between intensity and frequency of noise and the causal impacts between water and particles and between the particles themselves. The main result of this technique is that it can be successfully applied for the definition of the onset of particle mobilization and transport. There are some hydropower schemes that control the lowering of sediment exclusion weirs at their water intakes by using such microphones. However, there is still no simple, calibrated bed load-acoustic signal function for bed load noise detectors that can define the beginning and end of particle movement as well as the intensity of particle discharge. Nevertheless, it is anticipated that measurements of particle to particle impact, in addition to measurements of particle vibration, will one day create distinctive signals which are easier to detect than the noise of flood waters. It is the low frequencies that are of particular interest.

3.Tracers

Painted tracers and particles with exotic lithology represent the oldest approach to bed load transport measurement and according to Sear were first applied by Richardson in 1902.

He investigated the distribution of a pile of bricks on Chesil Beach in England. This interesting experiment nevertheless suffered from the very beginning due to the neglect of the following principles, i.e. that tracer particles should have a similar grain size, shape, specific gravity and erodibility as natural particles. During the last century the use of painted or tagged particles rapidly increased, since they are cost effective and capable of solving many different problems. More modem techniques were developed and were only applied during the last 20 years. In 1998, Foster convened the annual conference of the BGRG (British Geomorphological Research Group) on "Tracers in Geomorphology" and published the results in a special volume. The chapter contributed by Sear on "Coarse Sediment Tracing Technology in Littoral and Fluvial Environments" covers all important tracer methods and provides a good review of the literature. The new wave of tracer techniques gave rise to investigations of the following problems of coarse sediment transport:

- Boundary conditions or thresholds for initiation of motion.
- Rate and direction of bed load transport.
- Single particle transport: particle and bed properties and hydraulic conditions.
- Periods of movement and rest for single particles.
- Volumes of bed load transported and the magnitude of bed load discharge.

The new generation of coarse grain tracers comprise:

- Luminescent tracers.
- Radioactive tracers.
- Magnetic tag tracers.
- Magnetic and aluminum tracers.
- Radiotracers.

The performance and merits of these techniques will be discussed in the following sections.

4.Visual tracers

This type of tracer comprises all "old" tracer techniques and technically visible tracers created using luminescent dyes. The particles are coated with dye and are detected by the naked eye during the day (normal paint) or during the night (with the help of luminescent dye radiated by UV) if deposited at the surface. Since the dispersion is three-dimensional (3D), the recovery rate of particles in natural rivers after a flood wave is frequently only half that of the initial sample and, depending on the geomorphological setting, the recovery rates can be even worse. If only short particle transport distances are involved, as in the context of the construction and destruction of pebble clusters, or investigation of the potential for the initiation of particle motion or local morphological dynamics painted pebble techniques are still useful. Frequently, only a certain width of the river bed is activated during phases of bed load transport, and this can be effectively demonstrated using lines of painted particles .Nevertheless, the dominance of painted particles in bed load studies terminated at the beginning of the 1980s since there were too many shortcomings in the application of this technique for the determination of bed load discharge.

5.Radioactive tracers

Radioactive tracers were mostly used in the 1960s and 1970s for the study of sand transport both in rivers and along coasts. There is a vast specialized body of literature covering not only the British and French, but also the Eastern European experiences, in particular. Gravel and coarser material was mostly labelled by inserting "pills" containing radioactive elements. The most common radioactive isotopes for bed load movement are described in a table by Richards. The half-life should match the frequency of bed load transport of the river, although this principle is not always applicable. In Eastern Europe many experiments were executed with natural bed material activated in nuclear reactors. Dispersion is usually measured using a scintillation detector mounted on a sledge which is towed over the river bed. The measured spatial distribution can be interpreted using dispersion models. The most successful models are based on the Einstein approach, including assumptions of step-length distances and rest periods between periods of movement. These approaches were also applied to the results obtained from radio tracers 30 years later. Since the 1970s, tracer experiments using radioactive isotopes are no longer permitted in most western countries.

6.Magnetic tracers

Radioactive labelling of bed load was replaced by a new technique developed by the Berlin group of physical geographers at New Year 1980, in the River Buonamico in Calabria, southern Italy. A small magnetic bar was inserted into a hole drilled into a bed load particle. The cobbles were then placed into the bed of a small, steep creek feeding into the Lago Costantino. Bed load mobilization occurred due to snowmelt during the late afternoon of the first day of the experiment and the passage of tagged particles was successfully registered using the Faraday principle, involving a system of large coils (25 cm in diameter) mounted onto a metal beam suspended across the creek 30 cm above its bed. This first trial was so promising that it was demonstrated to the visiting Israeli scientists Schick and Hassan in summer 1980, during their excursion to southern Germany. From then on the Berlin group developed automatic magnetic measuring systems, whereas the Jerusalem group focused on the 3D dispersion of magnetically-traced particles. The resulting publication are numerous. The Berlin group cooperated with Montana State University in Bozeman, USA and installed their device at Squaw Creek, a small river whose bed load is predominantly andesitic containing a high amount of magnetic crystals. About 66% of all pebbles have a sufficient magnetic concentration to create signals above the electronic noise level. The technical potential of this approach, including the relation between signal volume and grain size, was highlighted by Spieker & Ergenzinger. The dissertations of Bunte and de Jong summarize the immense data set assembled by the studies undertaken by the Berlin group.

The magnetic tracer technique was applied to determine the movement of naturally or magnetically tagged particles via detectors on the one hand and via tagged nonmagnetic particles on the other hand, in order to trace their dispersion after floods with a special magnetometer. In contrast to the use of painted material, the magnetometer can sense the disturbance of the Earth's magnetic field approximately 30-50 cm below the surface of the sediment. As with radioactive tracers, it is possible to sense the dispersion of the material in three dimensions and to determine the volume of sediment moved by a single flood event from the average width of bed load streets (multiplied by the depth of the particle layer. In addition, magnetic tracers can be applied in many special fields. For example, Gintz manufactured hundreds of artificial magnetic particles with similar weights, but a variety of different shapes (ellipsoids, spheres, rods and disks), and used them to analyse the impact of particle shape on travel length.

The detector analysis is particularly well adapted for sensing the movement of natural magnetic particles. Magnetic particles cross a detector log and induce a microvolt current passing through the coil system. Movement of particles is detected with very high temporal resolution (in second intervals). The spatial resolution is defined according to the geometry of the coils and the number of coils defined over one field. The particle grain size can be derived with a high probability from the area of the induced signals. If several detector sections are mounted along the river, this makes it possible to sample the longitudinal transit of bed load. Under these conditions problems related to the movement of bed load particles in general can be solved.

7.Radio tracers

In the same month of the same year, the group of Emmett and of Ergenzinger developed the use of radio tracers for application in rivers. The American group worked in Alaska while the German group worked in Upper Bavaria. Radio tracers were selected to validate the Einstein model, by measuring the step length and rest periods of coarse single particles. For this purpose a micro transmitter, with a frequency of 150 MHz and batteries 55 mm in length and 20 mm in diameter, was installed in holes drilled into natural cobbles. With an antenna type HB 9 CV the signal of 1 mW could be received over a range of up to 200 m. For the determination of step length and rest periods, a band of 12 antennae was installed along the river bank to allow simultaneous observation of up to eight radio particles with a spatial accuracy of up to 2 m.

The technique proved to be successful in shallow waters with low conductivity. However, the applied frequency is not appropriate for more saline and deeper rivers such as the River Rhine, and is absolutely inappropriate for seawater. It is, nevertheless, surprising that the new device is not used more frequently in wadis with extreme flood peaks. There is a good chance of relocating radio-tagged cobbles, even months after insertion, and to define the dispersion of particles with different properties in families of 10-20 radio cobbles with different signals and similar or differing frequencies. The largest restriction is still the dimension of the battery. However, if the dispersion of particles is only relevant during the wet season, PETSY can be switched on after a defined time period. The signals can be located via helicopter or from vehicles. The average bed load discharge can be calculated nearly as precisely as water discharge under these conditions, from the median of the dispersal width and length and the average depth of the buried particles.

In cooperation with Christaller from the Technical College Berlin, a transmitter with different sensors, the telemetric device COSSY (Cobble Satellite System), was produced. The prototype was applied in a detailed study of the relationship between lift and drag forces under critical conditions. Similar investigations were undertaken by Mikos at the Technical University of Ljubljana, Croatia. Since radio tracing is very common nowadays, more devices for radio tracing are available and prices for ICs are decreasing. There is growing potential for more sophisticated devices for many special problems of bed load transport. For example, step and rest periods can be recorded and stored in individual cobbles that can be located and tapped after floods.

8. Conclusion and future

The geomorphology and engineering toolboxes contain a number of methods for measuring bed load transport in rivers. During the last 25 years, many new techniques have been proposed and applied. Due to the effects of environmental controls at the beginning of this period, the capability of radio-active tracers was lost but replaced by aluminium and magnetic tracers, and subsequently radio tracers. The most significant limitation of all these efforts is that there has been no joint experiments for validation purposes. It is especially important to carefully determine the minimum sample size required for radio tracers to assure the validity of results. Since these types of tracers are expensive, not more than 10 radio tacked particles are commonly applied and with this limit, the probability of the results matching natural transport is uncertain. In this context the development of new approaches, moving away from single particle or cross-sectional concepts, towards approaches linking the spatial distributions of river bed changes with related transport provides new opportunities. These in turn provide new opportunities for validating the results of small-scale experiments and model output. It is arguably surprising that during a new, important era where the significance of fluvial morphology has been increased due to the freshwater guidelines of the EU and other important bodies, the number of related experts involved is decreasing and interest in investigating field-based sediment transport is gradually diminishing in relation to modelling. This stands in clear contrast to the present-day requirements. At the same time, the potential to apply the results of local as well as regional investigations is growing rapidly.

New Words and Phrases

1.mount　*v.*　准备；安排　　　　2.sampler　*n.*　取样器

3.unaltered *adj.* 未改变的
4.ambient *adj.* 周围的
5.lengthy *adj.* 很长的；漫长的
6.acoustic detector *n.* 声探测器
7.hydro-microphone *n.* 水听器
8.collision *n.* 碰撞
9.onset *n.* 开端，发生
10.exclusion *n.* 排斥；排除在外
11.tracer *n.* 示踪剂
12.initiation *n.* 开始；创始；发起
13.luminescent *adj.* 发冷光，冷光的
14.dominance *n.* 支配；控制
15.radioactive isotope 放射性同位素
16.magnetic tracer 磁性示踪剂
17.immense *adj.* 极大的；巨大的
18.ellipsoid *n.* 椭圆面，椭球面
19.sphere *n.* 球；球体
20.rod *n.* 杆；竿
21.disk *n.* 磁盘；磁碟
22.wadis *n.* （中东和北非仅在雨后才有水的）干谷，干河谷
23.helicopter *n.* 直升机
24.arguably *adv.* 可论证地，按理

Notes

1.There is a continually growing list of important publications dedicated to bed load functions, but data for validation is spare and mostly restricted to flume experiments.

目前，有一系列不断增长的出版物主要注重于推移质运动的方程，而缺乏实测验证数据和水槽试验数据的限制。

dedicated to 把（时间、力量等）用在…… be dedicated to 致力于；献身于

A monument was dedicated to the memory of the national hero.

2.This interesting experiment nevertheless suffered from the very beginning due to the neglect of the following principles, i.e. that tracer particles should have a similar grain size, shape, specific gravity and erodibility as natural particles.

然后，这个有趣的试验在最初有一定的困难，主要是由于忽略了以下因素：示踪颗粒具有与天然颗粒类似的粒径形状、相对密度和侵蚀度。

nevertheless 尽管如此；不过；然而。是指即使做出完全的让步，也没有任何影响，表达的深度比 however 更深

Nevertheless, available data, even if insufficient, together with scientific theory make it possible to solve a whole set of water balance problems as well as to estimate world water balance, provided suitable methods for investigating are applied.

3. In this context the development of new approaches, moving away from single particle or cross-sectional concepts, towards approaches linking the spatial distributions of river bed changes with related transport provides new opportunities.

本文中所提出的新方法，摆脱了单颗粒或者断面的概念，为河床空间分布与相关运输过程研究纽带提供了新思路。

move away from 本身在某个地方或靠近某个地方，但要离开这个地方

This approach was concomitant with the move away from relying solely on official records.

Comprehensive Exercises

I.Answer the following questions on the text.

1.How many methods we can measure the sediment transport?

2.What is the limitation of the sampler?

3.What is the characteristics of the Acoustic detector?

4.Why is the Radio tracer not appropriate for more saline and deeper rivers?

5.How do we improve the knowledge of sediment behavior in rivers?

II.Fill the most appropriate words or phrases in the correct forms in the blanks from the list below.

collision	due to	capable of	Initiation	In contrast to
distribution	spatial	suffer from	in turn	in cooperation with

1.However, the understanding of river-runoff ____over the area has substantially improved.

2.Ice jam and ice dam often occur ______time difference of freeze-up and melting between upstream and downstream in winter and spring.

3.In ______oil pipeline divulging accident all sorts of reasons, the artificial destruction pipeline steals the petroleum the situation to be most serious.

4.The ______ extent of the impact varied among sites.

5.Rotary action of the turbine ______drives an electrical generator that produces electrical energy or could drive other rotating machinery.

6.Christaller from the Technical College Berlin, a transmitter with different sensors, the telemetric device COSSY (Cobble Satellite System), was produced.

7.Refugees settling in Britain _______a number of problems.

8.In flat angle_________, most vehicles are redirected by the "safety shape" with no damage and are able to drive away.

9.Reciprocating pumps, sometimes called piston or displacement pumps______developing high heads, but their capacity is relatively small.

10.______these rapid-recharge aquifers, other aquifers allow very slow water flow, perhaps a few hundred meters per year.

III.Translate the following sentences into Chinese from the text.

1.Since the beginning of construction of hydropower schemes, the demand for improved information on bed material transport has increased. A bag was mounted onto a frame, or sediment was trapped in mesh-covered boxes suspended from bridges. For finer particles, such as sand and gravel, these approaches were replaced by pressure difference samplers.

2.If only short particle transport distances are involved, as in the context of the construction and destruction of pebble clusters, or investigation of the potential for the initiation of particle motion or local morphological dynamics painted pebble techniques are still useful.

3.The technique proved to be successful in shallow waters with low conductivity. However, the applied frequency is not appropriate for more saline and deeper rivers such as the River Rhine, and is absolutely inappropriate for seawater. It is, nevertheless, surprising that the new device is not used more frequently in wadis with extreme flood peaks

4.It is arguably surprising that during a new, important era where the significance of fluvial morphology has been increased due to the freshwater guidelines of the EU and other important bodies, the number of related experts involved is decreasing and interest in investigating field-based sediment transport is gradually diminishing in relation to modelling.

Reading Material Flume

Of the two primary devices for open-channel flow measurements, flumes are preferred to weirs on account of following points:

1) The water must be dammed for flow measurements by weir. This may affect the inflow region.

2) On account of approach velocity effects, a large upstream stilling basin is required for weirs.

3) In weirs there may be deposits of silt and other solids that might accumulate and alter the measurements.

4) Energy loss is smaller compared to that in a weir.

Open-channel flumes (applicable to weirs also) are designed in a manner to force a transition from subcritical to supercritical flow. For flumes, the transition is caused by constricting the side walls of flumes to have a narrowing at the throat, raising of the channel bottom, or both. Such construction causes flow to pass through a critical depth at the flume throat. As has already been seen, at the critical depth, energy is minimized as discharge/velocity holds a direct relationship with water depth. However, in flumes it is not easy to measure the critical depth as its exact location is difficult to determine with the varying flow rate. As discussed above, through mass conservation, the upstream depth is related to the critical depth so that discharge can be determined by measuring the reliable upstream depth.

1.Flume classes

The primary device flume is classified below. For better understanding of the functionality of different types of flumes.

1) Long-throated flumes: Long-throated flumes have too long control discharge at the prismatic throat section, with sufficient length in the streamwise direction to achieve a nearly parallel flow situation and a hydrostatic pressure distribution. Flumes can be rated through proper analysis using fluid flow concepts. The Palmere Bowlus is a long-throated flume.

2) Short-throated flumes: These flumes are named short-throated flumes because they control flow in a region of curvilinear flow. A short-throated flume is characterized by a strong free surface curvature and a departure from the hydrostatic distribution of pressure. The Parshall flume is the most popular short- throated flume. The overall dimensions of the flume are not too short despite being called a short-throated flume. Calibrations for short-throated flumes are determined empirically by comparison.

3) Special flumes: H-flume: H-flumes are made of simple trapezoidal flat surfaces. These surfaces are placed in such a way as to form vertical converging sidewalls. The downstream edges of the trapezoidal sides are placed so that they form a notch, widening with distance from the bottom. The maximum allowed submergence is 30%.PalmereBowles flumes: Palmere Bowles flumes are constructed as inserts with circular bottoms to fit conveniently into U-shaped channels or partially full pipes. These flumes are of the long-throated type. Cutthroat flumes: Cutthroat flumes are Par-shall flumes with the throat "cut out", hence the name. These flumes are formed by joining a 6:1 converging section with a similar diverging section. They do not have any parallel walls to form a throat. This is a simplified version of the Parshall flume.

Flat-bottomed trapezoidal flumes: These are flat-bottomed trapezoidal short-form flumes. They were designed to sit flush with respect to the bottom of the incoming channels in an effort to assist sediment movement and allow the canal to drain dry between uses.

Special flumes for passing sediment: These specially designed flumes are designed to combat sedimentation problems, and measure the flow in an open channel. They are super-critical flumes, requiring extensive head drop to operate. They have very limited applications in irrigation.

2.Explanation of flume geometry

Normally, flumes have three sections. In the inlet converging section the liquid is forced to accelerate. Liquid enters the converging section in a subcritical state, i.e., typically with F_r around 0.5. The liquid head is normally measured at this section (for submergence it acts as ha discussed earlier). Liquid acceleration takes place mainly at the throat and flow goes to the critical and supercritical stages, $F_r \leqslant 1$. The next section is the divergent section which slows down the energetic flow to allow the same to the downstream channel. Naturally, where the divergent section is absent the flow will be susceptible to erosion/scour.

3.Pros and cons of flumes

Listed below are the major advantages and disadvantages of flumes. Advantages: Following points are a few advantages of flumes.

1) Flumes do not require a dam across the flow line or upstream stilling chamber.

2) They are self-cleaning on account of the higher velocity.

3) For same flow rates flumes have much lower head loss.

4) Flumes offer a wide selection of flow measurement, conditioning, and control options.

5) Flumes offer also a wide selection of cross- sections and shapes.

6) Flumes offer moderate accuracy but give dependable and repeatable measurements.

There are no sharp edges/pockets and not too many critical dimensions in flumes. Disadvantages: There a few disadvantages of flume also and these are:

A major limitation is the cost of fabrication and installation.

Accuracy is highly dependable on the style and type of flume.

Flow equations are a little more complicated, especially for certain types, e.g., H type flumes.

Lesson 3 Numerical Model

Numerous river engineering problems can be conveniently investigated by means of mathematical models. Mathematical models must properly describe the physical processes and provide a numerical solution to a system of differential equations that are solved together with suitable boundary conditions and empirical relationships that describe resistance to flow and turbulence.

The differential equations describing river mechanics problems are usually simplified forms of the equations of conservation of mass and momentum, leading to a set of partial differential equations involving two independent variables (time and space or two spatial variables).The finite-element method also provides useful solutions to river engineering problems but is beyond the scope of this text.

The algorithms to be used in the finite-difference method depend on the type of differential equation to be solved. Table 4.1 provides a simple classification of river engineering problems. The information propagates at a celerity c in hyperbolic equations, and the celerity is effectively infinite in parabolic equations.

Table 4.1 **Differential equation types in river engineering**

Equation type	Equation	Equation type problem
Hyperbolic	$\frac{\partial \phi}{\partial t}+v\frac{\partial \phi}{\partial x}=0$	Advection(v constant)
	$\frac{\partial^2 \phi}{\partial t^2}=c^2\frac{\partial^2 \phi}{\partial x^2}$	Floodwave propagation(c^2 constant)
Parabolic	$\frac{\partial \phi}{\partial t}=K_d\frac{\partial^2 \phi}{\partial x^2}$	Diffusion-dispersion(K_d constant)
Elliptic	$\frac{\partial^2 \phi}{\partial x^2}+\frac{\partial^2 \phi}{\partial y^2}=0$	Flow net

Once a river engineering problem has been defined and a mathematical model chosen, field data need to be gathered to describe initial and boundary conditions, geometrical similitude, material properties, and design conditions. Additional data are also required for calibration and verification. The governing equations can be simplified to preserve the main features of the physical problem; the time and the space increments are determined at this stage. A schematization can be made of the design conditions to be investigated.

Model calibration is usually necessary because empirical parameters are involved to describe resistance to flow and because of the implifications to the governing equations. Parameters can be adjusted to obtain good correspondence between numerical results and continuum values. Of course, the adjustment should not be extended beyond physically acceptable values. The precision of a model refers to the error margin of the numerical calculations. Model accuracy usually refers to the comparison of the model with field measurements. For instance, a model that calculates the floodstage to the nearest centimeter but is 1 m off from the field measurements is precise but not accurate. The method of adjusting parameters by running the model at different values until a satisfactory result is obtained is called hindcasting. It is a very useful way to determine the sensitivity of the model results to changes in the model parameters. The calibration phase should also comprise a check of the numerical accuracy by varying numerical parameters such as the time step.

Model verification involves simulation for a different set of prototype data with the coefficients previously obtained during the calibration. If a model run satisfactorily reproduces the measured prototype conditions without further adjustment, a reasonable confidence is gained in the application of the model to design conditions that have never occurred in the prototype. It is often possible to calibrate a model with the first half of a field data set and to verify the model with the second half.

1. Finite-difference approximations

Let us consider a function $h(x, t)$ defined in space x and time t. We may divide the x-t plane into a grid, as shown in Fig. 4.1(a). The grid spacing along the x axis is and the time interval along the t axis is Δt.

The value of the variable h will use the spatial location as a subscript and the time as a superscript, e.g., h_k refers to the value of flow depth at the jth spatial grid point and kth time grid point. By the known time level, we mean that the values of different dependent variables are known at the time level h_k and we want to compute their values at the unknown time level h_{k+1}. If the computations progress from one step to the next, then the procedure is referred to as a marching procedure. Most of the phenomena described by hyperbolic partial differential equations are solved with marching procedures. The conditions specified at time $t = 0$ are referred to as the initial conditions. The conditions specified at the channel ends are called the end, or boundary, conditions.

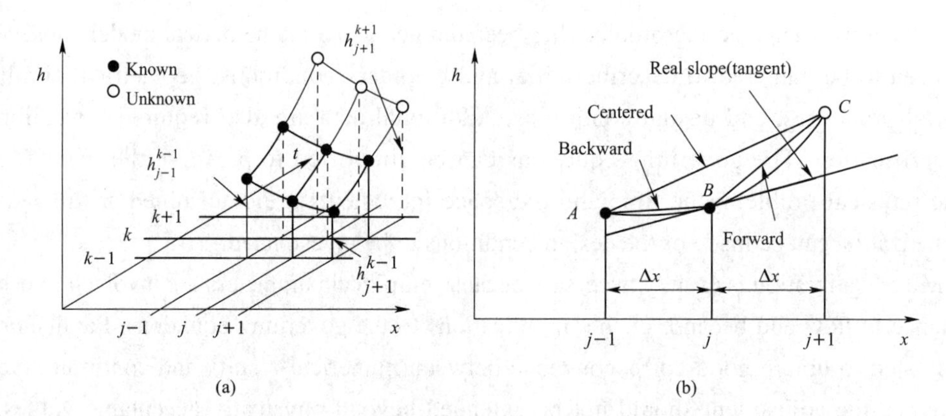

Figure 4.1 Finite-difference grid and approximation.

Finite-difference approximations are first introduced before a presentation on consistency and convergence, a linear stability analysis, higher-order approximations, and boundary conditions. The finite-difference method is based on a Taylor series expansion of the variable h_{j+1} written as a function of h_j as:

$$h_{j+1}^{k}=h_{j}^{k}+\Delta x\left(\frac{\partial h}{\partial x}\right)_{j}^{k}+\frac{\Delta x^{2}}{2!}\left(\frac{\partial^{2} h}{\partial x^{2}}\right)_{j}^{k}+\frac{\Delta x^{3}}{3!}\left(\frac{\partial^{3} h}{\partial x^{3}}\right)_{j}^{k}+0\left(\Delta x^{4}\right) \tag{4.3}$$

where the derivative $(\partial h/\partial x)_j^k$ is evaluated at grid point j and time level k and $0(\Delta x^m)$ indicates m=order terms. The Taylor series could be similarly expanded to define h_{j-1}^k from h_j^k as:

$$h_{j-1}^{k}=h_{j}^{k}-\Delta x\left(\frac{\partial h}{\partial x}\right)_{j}^{k}+\frac{\Delta x^{2}}{2!}\left(\frac{\partial^{2} h}{\partial x^{2}}\right)_{j}^{k}-\frac{\Delta x^{3}}{3!}\left(\frac{\partial^{3} h}{\partial x^{3}}\right)_{j}^{k}+0\left(\Delta x^{4}\right) \tag{4.4}$$

Rearranging Eqs. (4.3) and (4.4) and dividing by Δx gives, respectively,

$$\left(\frac{\partial h}{\partial x}\right)_{j}^{k}=\frac{h_{j+1}^{k}-h_{j}^{k}}{\Delta x}-\frac{\Delta x}{2!}\left(\frac{\partial^{2} h}{\partial x^{2}}\right)_{j}^{k}-\frac{\Delta x}{3!}\left(\frac{\partial^{3} h}{\partial x^{3}}\right)_{j}^{k}+0\left(\Delta x^{3}\right)\cong\frac{h_{j+1}^{k}-h_{j}^{k}}{\Delta x}+0(\Delta x) \tag{4.5}$$

$$\left(\frac{\partial h}{\partial x}\right)_{j}^{k}=\frac{h_{j}^{k}-h_{j-1}^{k}}{\Delta x}-\frac{\Delta x}{2!}\left(\frac{\partial^{2} h}{\partial x^{2}}\right)_{j}^{k}-\frac{\Delta x}{3!}\left(\frac{\partial^{3} h}{\partial x^{3}}\right)_{j}^{k}+0\left(\Delta x^{3}\right)\cong\frac{h_{j}^{k}-h_{j-1}^{k}}{\Delta x}+0(\Delta x) \tag{4.6}$$

The first approximation of the partial derivative in Eq. (4.5) is written in the form of a forward (downwind) difference and a first-order truncation error $0(\Delta x)$ that will approach zero as Δx becomes very small. Similarly, Eq. (1.4) includes a backward (upwind) difference and a truncation error $0(\Delta x)$. The truncation error approaches zero as Δx approaches zero as long as the high derivatives remain continuous. Therefore both forward and backward finite-differences are first-order approximations.

A central finite-difference approximation can then be obtained from taking half of the sum of Eqs. (4.5) and (4.6), or

$$\left(\frac{\partial h}{\partial x}\right)_j^k = \frac{h_{j+1}^k - h_{j-1}^k}{2\Delta x} + 0\left(\Delta x^2\right) \tag{4.7}$$

The truncation error is of the order of (Δx)^2 because the terms in Δx in Eqs. (4.5) and (4.6) cancel. The central difference is thus said to be a second-order approximation.

Fig. 4.1(b) shows a geometrical representation of the forward, backward, and central finite-difference approximations. The real slope is the tangent of the function at B. The forward finite-difference approximation uses the slope of the secant curve line BC, the backward finite-difference approximation uses the slope of line AB, and the central finite-difference approximation uses the slope of the line AC, although all three approximations become exact as goes to zero (first-order approximation). It is clear from these figures that the central finite-difference approximation is more accurate (second-order approximation) than the forward or the backward finite- difference approximations.

Explicit formulations refer to partial derivatives at the known level k whereas implicit formulations refer to the unknown level k+1. Table 4.2 lists some typical explicit and implicit finite-difference approximations for the spatial partial derivative, $\partial h/\partial x$, at the grid point (j,k).

Table 4.2 **Explicit and implicit finite differences**

Finite difference	Explicit	Implicit
Backward upwind	$\frac{\partial h}{\partial x} \approx \frac{h_j^k - h_{j-1}^k}{\Delta x}$	$\frac{\partial h}{\partial x} \approx \frac{h_j^{k+1} - h_{j-1}^{k+1}}{\Delta x}$
Forward downwind	$\frac{\partial h}{\partial x} \approx \frac{h_{j+1}^k - h_j^k}{\Delta x}$	$\frac{\partial h}{\partial x} \approx \frac{h_{j+1}^{k+1} - h_j^{k+1}}{\Delta x}$
Central	$\frac{\partial h}{\partial x} \approx \frac{h_{j+1}^k - h_{j-1}^k}{2\Delta x}$	$\frac{\partial h}{\partial x} \approx \frac{h_{j+1}^{k+1} - h_{j-1}^{k+1}}{2\Delta x}$

2.Consistency and convergence

Four properties of consistency, stability, convergence, and accuracy are important in numerical analysis. The following formulation of the advection equation, or floodwave propagation problem, is used to illustrate these properties. Hence

$$\frac{\partial h}{\partial x} + c\frac{\partial h}{\partial x} = 0 \tag{4.8}$$

where c is the celerity, is approximated with a forward difference in time and a backward difference in space (FTBS) to give

$$\frac{h_j^{k+1} - h_j^k}{\Delta t} + c\frac{h_j^k - h_{j-1}^k}{\Delta x} = 0 \tag{4.9}$$

Rearranging to find the flow depth at the unknown level k+1 as a function of the flow depth at the known level k, we obtain

$$h_j^{k+1} = h_j^k - \frac{c\Delta t}{\Delta x}\left(h_j^k - h_{j-1}^k\right) \tag{4.10a}$$

$$h_j^{k+1} = C_c h_{j-1}^k + \left(1 - C_c\right) h_j^k \tag{4.10b}$$

where $C_c = (c\Delta t / \Delta x)$ is the Courant number.

To get started, the initial condition of flow depth needs to be known for all j values at $k = 0$. The algorithm can then march in time given the boundary condition of flow depth for all k values at $j-1 = 0$.

Consistency is the property of a finite-difference scheme to reduce to the partial differential equation as the truncation error disappears. In our example, the values h_{j+1}^k and h_{j-1}^k from Eqs. (4.3) and (4.4) are substituted back into Eq. (4.10b) to give

$$h_j^k + \Delta t\left(\frac{\partial h}{\partial x}\right)_j^k + \frac{\Delta t^2}{2!}\left(\frac{\partial^2 h}{\partial x^2}\right)_j^k + 0\left(\Delta t^3\right) = \left(1 - \frac{c\Delta t}{\Delta x}\right)h_j^k + \frac{c\Delta t}{\Delta x}\left[h_j^k - \Delta x\left(\frac{\partial h}{\partial x}\right)_j^k + \frac{\Delta x^2}{2!}\left(\frac{\partial^2 h}{\partial x}\right)_j^k + 0\left(\Delta x^3\right)\right] \tag{4.11}$$

A method is said to be convergent when the difference between the solutions of the differential and difference equations tends to zero as the time step goes to zero. It has been shown that a consistent method, if stable, is also convergent and vice versa. Consequently it is generally sufficient to check consistency and stability to ensure convergence. It is therefore indicated to examine the stability of numerical schemes.

3. Linear stability analysis

The stability of a difference method is concerned with the propagation of an error, introduced for example by inaccurate initial or boundary data or rounding in the numerical calculations. Such errors will be propagated by the difference method. If they do not grow, the method is called stable.

The linear stability analysis, also referred to as the von Neumann procedure, examines the property of the response of the finite-difference scheme to input perturbations written as a Fourier series in complex form as the Courant- Friedrich-Levy (CFL):

$$C_c = \frac{c}{\Delta x / \Delta t} \leqslant 1 \tag{4.12}$$

The physical interpretation of the CFL condition is that the numerical time step may not exceed the characteristic time step $\Delta x/c$. Otherwise, all the physical information does not have sufficient time to propagate to the next time step, and this will manifest itself as an instability. In practice, the CFL condition often imposes an important restriction on the time step because the spatial step Δx is generally determined by geometric considerations. This restriction has led many investigators to prefer implicit methods for several decades. With the development of very fast computers, explicit numerical schemes are nowadays gaining popularity.

4. Boundary conditions

Elliptic equations require that boundary conditions be specified over a completely closed boundary. The boundary data may consist of the value (Dirichlet type of boundary condition) or its normal derivative (Neumann type of boundary condition). The solution to hyperbolic and parabolic equations usually requires both initial and boundary conditions. Starting from given initial conditions at t=0, a marching method finds the values at successive unknown time intervals from the boundary conditions.

The zone of influence of boundary conditions propagates through the domain at each time step. Lower-order approximations require simple boundary conditions. For instance, a scheme $h_j^{k+1}=0.5h_{j-1}^k+0.5h_j^k$ requires only the upstream boundary condition h_0^k and the initial condition h_0^j for calculations over the entire domain. A downstream boundary condition is not required.

Besides the initial condition at all nodes h_0, this algorithm requires a double upstream boundary condition at h_{j-1} and h_0 and also requires a single downstream boundary condition because the term h_{j+1} implies that the information propagates in the upstream direction. The term h_{j+1} also implies that the information contained at the upstream boundary condition does not entirely propagate in the downstream direction. Indeed, some information contained in the upstream boundary condition propagates outside of the solution domain, in the upstream direction. This becomes particularly important in advection-dispersion problems in which conservation of mass is expected. Higher-order approximations thus provide improved calculation results at the expense of requiring detailed boundary conditions.

New Words and Phrase

1.mathematical *adj.* 数学的
2.algorithms *n.* 算法
3.celerity *n.* 运动的速度，波速
4.hyperbolic *adj.* 双曲线的
5.infinite *adj.* 无限的
6.parabolic *adj.* 抛物线的
7.elliptic *adj.* 椭圆的
8.similitude *n.* 类似，相似
9.preserve *v.* 保护，保留
10.increment *n.* 增量
11.schematization *n.* 详略化，规范化
12.implification *n.* 简化
13.hindcaste 追算
14.prototype *n.* 原型，雏形
15.consistency *n.* 一致性，连贯性
16.convergence *n.* 融合
17.approximation *n.* 近似值
18.expansion *n.* 扩展
19.derivation *n.* 起源，来源
20.truncation *n.* 截除；切断
21.backward *adj.* 向后的
22.upwind *adj.* 迎风的
23.secant *n.* 正割
24.explicit *adj.* 明确的
25.implicit *adj.* 内含的， 完全的

Notes

1.The differential equations describing river mechanics problems are usually simplified forms of the equations of conservation of mass and momentum, leading to a set of partial differential equations involving two independent variables (time and space or two spatial variables).

描述河流机理问题的微分方程通常为简化的连续方程与动量方程，因而其中部分微分方程涉及两个独立变量（时间空间或者两者）。

lead to　引起，后面通常接结果

the subsidence of the Nile Delta will lead to inundation of the northern portion of the delta with seawater, in areas which are now used for rice crops.

a set of　一套，主句的谓语动词应该为单数

Each option has a set of design criteria for location within the landscape, to reduce risk, improve planning and provide maximum economic benefit to the landholder.

2.Once a river engineering problem has been defined and a mathematical model chosen, field data need to be gathered to describe initial and boundary conditions, geometrical similitude, material properties, and design conditions.

一旦河流工程问题提出并选择了相应的数学模型，现场测量数据需要被收集起来以便用于描述初始和边界条件、地貌特性、床面物质属性以及设计工况。

once　一旦……就；若从句中的主语与主句主语一致，可以省略从句主语。

once the project is completed in 2009 the depth of normal level of the water in the reservoir will be 175 meters.

3.The stability of a difference method is concerned with the propagation of an error, introduced for example by inaccurate initial or boundary data or rounding in the numerical calculations.

有限差分计算的稳定性与累计误差有关，该误差主要是由于初始或者边界条件不准确或者是数值计算中的凑整引起。

rounding　凑整　　be concerned with　与某事物有关

Construction engineering is a specialized branch of civil engineering concerned with the planning, execution, and control of construction operations for such projects as highways, buildings, dams, airports, and utility lines.

Comprehensive Exercises

I.Answer the following questions on the text.

1.How many types of differential equation in river engineering?

2.What does involve in the model verification?

3.What is the boundary conditions in the numerical model?

4.How to distinguish the stability of the numerical model?

5.How to improve the precision of the numerical model?

II.Fill the most appropriate words or phrases in the correct forms in the blanks from the list below.

a set of	beyond	infinite	in the form of	refer to
concerned with	stability	interpretation	imply	propagate

1.The outlet channel conveys and returns the water to the stream _____the dam or into other topographic depressions beyond the reservoir basin.

2.It is understood, of course, that there are ______variations on the theme.

3.Engineering______actions to be taken in the future, an important part of the engineering process is improving the certainty of decisions with respect to satisfying the objectives of engineering applications.

4.One method of laying a reinforced concrete wearing surface is put down the steel roads, usually _______a grill or mesh, after a certain proportion of the concrete has been poured.

5.Gravity dams ______solid concrete or masonry dams of roughly triangular cross section, which depend primarily on their own weight and cohesion with the foundation for stability.

6.The ______of the word 'needs' should go beyond physical needs of industry and agriculture, into the realm of spiritual needs.

7.A saturation condition ______that the moisture contents typically will be in the range of 40%-60%.

8.The information ______at a celerity c in hyperbolic equations, and the celerity is effectively infinite in parabolic equations.

9.Some authorities ______classify these pumps separately from ordinary centrifugal because all impellers do not function entirely on the centrifugal principle.

10.______guidance notes is provided to assist applicants in completing the form.

III.Translate the following sentences into Chinese from the text.

1.Mathematical models must properly describe the physical processes and provide a numerical solution to a system of differential equations that are solved together with suitable boundary conditions and empirical relationships that describe resistance to flow and turbulence.

2.The real slope is the tangent of the function at *B*. The forward finite-difference approximation uses the slope of the secant curve line *BC*, the backward finite-difference approximation uses the slope of line *AB*, and the central finite-difference approximation uses the slope of the line *AC*, although all three approximations become exact as goes to zero (first-order approximation).

3.Indeed, some information contained in the upstream boundary condition propagates outside of the solution domain, in the upstream direction. This becomes particularly important in advection-dispersion problems in which conservation of mass is expected. Higher-order approximations thus provide improved calculation results at the expense of requiring detailed boundary conditions.

4.For 1D flow, two alternatives of this scheme are possible: (1) backward finite differences are used in the predictor part and forward differences are in the corrector part; or (2) forward finite differences are used in the predictor part and backward differences in the corrector part.

Reading Material Multidimensional River Models

2D depth-integrated models have been applied to predict surface runoff and sediment-transport rates. A quasi-steady approach can often be used, although some 2D unsteady-flow models are available. A complete description of multidimensional models is beyond the scope of this chapter. A significant number of 2D and 3D codes are commercially available, and some are readily available in the public domain. The fast development of computers make new numerical solutions possible to river engineering problems of increasing complexity. In many cases in which the vertical variation in flow velocity and turbulence are of little interest, vertically averaged horizontal 2D models can be used.

Quasi-3D flow models simulate depth-averaged mathematical flow models in combination with the depth-integrated velocity profile. The continuity and the momentum equations are solved with empirical logarithmic velocity profiles. The bedload transport rate in the transverse direction is calculated with empirical formulas. The depth-averaged steady-state equation for the suspended load is described in an orthogonal, curvilinear coordinate system. Models can solve the sediment continuity equation for the 2D bed and suspended sediment transport. Once the depth-averaged 2D velocity calculations are completed, standard logarithmic velocity profiles are considered to determine the vertical velocity profile. Because the model calculations are in two dimensions, secondary currents in river bends cannot be properly simulated with quasi-3D models.

3D models are generally steady-state models used for turbulent-flow simulations. In κ–ε models, the state of turbulence is characterized by the energy and dissipation parameters κ and ε. 3D models typically solve the depth-averaged Reynolds approximation of the momentum equation for velocity. The depth- averaged mass conservation determines the water-surface elevation. The deviation from the depth-averaged velocity is computed for each cell by the solution of the conservation of mass equation in conjunction with a κ–ε closure for vertical momentum diffusion. Sedimentation computations are based on 2D solid mass conservation for the channel bed and the exchange of sediment between bedload and suspended load.

Data required for running multidimensional models include: (1) channel geometry with cross sections; (2) upstream/downstream boundary conditions in terms of discharge and stage as functions of time for unsteady-flow models and flow-velocity profiles for 3D models;

(3) particle-size distribution of the bed material; (4) upstream/downstream sediment load (some models require both bedload and suspended load); and (5) suspended sediment concentration profiles for 3D models.

The data requirements increase with the number of dimensions in the model. Some model features may require data that are not available, and many times assumptions must be made regarding missing data. For instance, some models can calculate sediment transport by size fractions, and sediment data of the bed material and boundary conditions may not be available for each size fraction.

Steady 3D models are applied to estimate the initial rate of sedimentation and erosion in a given situation. The reason for this is the vast computer time required for stabilizing the models under steady-state conditions. To run the model over long time periods under different flow conditions to determine aggradation and degradation would be prohibitive. The initial models provide good insight into the short-term effects of a proposed structure (channel diversion, new harbor, closure of a channel, etc.). However, they are of limited value for long-term simulation. It is usually preferable to run long-term 1D simulations in parallel to gain basic knowledge of morphological processes and long-term changes to be expected at the site.

Lesson 4 GIS and Big Data

Data are everywhere—in all shapes and sizes, coming from all directions and knocking persistently at our door. The age of information is providing mountains of ever-growing digital data—some data are managed and stored for examination and study; some are easily lost or forgotten. We try to reason that we don't need data from the past; we can always collect more in the future. However, as with history, if we forget or disregard data from the past, we doom ourselves to repeat analyses, waste resources, and squander time.

Somehow, we continue to collect data. Even through lean economic times, we march on and store and post our data for others to see and use. Through the Internet, various entities both public and private provide easily accessible data at little or no cost. We know, just as our ancestors did when they passed down information through oral storytelling or the recording of events on stone or parchment, that the information we collect and maintain today may be useful for future generations. We must remember, however, to examine our data and not just amass and store it.

1.Assessing spatial data using geographic information systems

Geographic Information Systems (GIS) offer a way to look at data in a spatial context. We abstract or simplify the real world into two types of spatial conceptualizations and corresponding models: the discrete view of objects in space (corresponding to vector representations) and the continuous view of those objects (analogous to raster representations). Vector models use points, lines, and polygons to respectively represent objects such as trees, transmission lines, and parcels of land. These geometric objects can contain a plethora of associated characteristics that in the vector model are traditionally stored as records in tables. Traditional tabular data can then be linked to these geometric features so that common database functions, such as queries and statistical summaries, can be seen graphically on a map.

Vector representations are appropriate for data sets having an explicit spatial location. Roads, for example, are better represented as discrete vector data because they have an explicit location and existence; they rarely taper off into nothingness. Additionally, most people do not care where roads are not; they only want to know the fastest and easiest way to pick up groceries and drop off the kids.

Thematic raster data models further abstract the real world into continuous arrays of cells or pixels. Each cell contains a discrete value demarcating whether or not (or how much of) the objects or phenomena of interest exist at a given location. A thematic raster representation of data may be more appropriate in situations where all locations within a spatial extent have a discrete meaning, such as a land cover data layer where each pixel in the array has a discrete classification category. A good example of a continuous raster representation is a terrain model, where each pixel represents a vertical elevation value. Photography, satellite imagery, and the like are also examples of continuous raster representations in which the pixel has a value. However, these pixels may not have discrete meaning until further analyzed using image processing techniques. Multi-spectral raster representations are simply a stack of raster representations, with each layer representing a discrete range of spectrum or other calculated values.

Many years ago, there was great debate in the GIS realm as to whether vector or raster representations of data were better. This debate has virtually disappeared because GIS analysts and data managers now understand that there is a technical place for both and that, depending on the data being represented and the task at hand, one format may better represent the information.

2.Data management, GIS, and sustainability

Technology has made an undeniable impact on our world. It has improved our health, expanded our population, and maximized our use of available resources. However, these same improvements are also detriments in terms of the health of our planet. So far, we have not achieved a sustainable balance. GIS and sound data management are important components of sustainability. Whether they are used to justify socioeconomic decisions, determine the feasibility of renewable energy sources, or assess environmental degradation, GIS and data management will play a key role in shaping our world today and in the future. The following sections discuss such applications.

We have become an energy-dependent society. Energy has a place in every facet of our lives, and its use is growing exponentially, especially in developing countries like India and

China. Large and reliable amounts of energy are required to fuel global development and growth. However, energy derived from fossil fuels is undeniably finite and has been shown to be harmful to the environment during harvesting, manufacturing, and use. Renewable energy (RE) is a long-term, sustainable solution. GIS can also be used to further RE technologies. For example, we can use GIS to determine where to construct biomass power plants with reliable fuel sources or the optimal placement of wind turbines. To accomplish such analyses, multiple spatial data sets must be combined using specialized analysis tools. First, for thorough and accurate answers, we need the appropriate data represented with the appropriate data model. Second, we need an understanding of the tools and principles of data analysis. Third, we need the ability to represent and communicate the results of our analysis. A brief example of how GIS was used to identify the location for potential biomass projects in Georgia is discussed in the following paragraphs.

3.Determining the location-specific potential for biomass projects

According to the National Renewable Energy Laboratory (NREL, n.d.), biomass is plant matter such as trees, grasses, agricultural crops, or other biological material. It can be used for fuel in its solid form or can be converted into liquid or gas and used to produce fuels, electricity, heat, and chemicals. Globally, biomass is currently the fourth largest producer of energy behind classic fossil fuels such as oil, coal, and natural gas. Researchers estimate that there are 278 quadrillion BTUs of installed biomass capacity worldwide (NREL, n.d.). With such a wide variety of possible sources and great potential over such a large area, tools and techniques are needed to locate biomass resources and possible production facilities efficiently and effectively.

The state of Georgia is a prime candidate for biomass development, with approximately 23000000 acres that could qualify as biomass stock. Many tools and techniques were available for resource and facility site identification in the state. This case study presents one technique: an additive model used to combine environmental and infrastructure parameters to determine locations that had viable amounts of biomass resources and that were suitable for biomass energy facility construction.

1) Data acquisition

The first step was to identify data needs. Discussions with foresters and research into biomass identified the following critical data sets:

Land cover type. Land cover types, and proximity to certain land covers, in many ways dictate biomass potential and a location's potential as a biomass facility site. Forested areas and agricultural lands provide more biomass feedstock than, say, an urban environment. Land cover data from the National Land Cover Database consisting of 29 land cover classifications was used for this project. It was recoded into ten classes of interest by combining similar class types (e.g., medium and heavy urban).

Percent tree canopy. Percent tree canopy data was used to quantify biomass and open space availability. Areas with a higher percent of tree canopy likely have elevated amounts of biomass in accordance with accepted biomass definitions. Conversely, areas with a lower percent of canopy and adjacent to a higher percent of canopy are better for biomass facility construction because of their available resources.

Biomass resource potential. Biomass approximations by county were used to quantify seven biomass feedstock categories: crop residues, methane emissions from manure management, methane emissions from landfills and wastewater treatment facilities, forest residues, primary and secondary mill residues, urban wood waste, and dedicated energy crops (NREL, n.d.). Because these data are at the county level, it gives general locations for potential biomass projects but nothing more granular. It was created using methods accepted by the NREL.

Proximity to high-voltage power lines. As one of many infrastructure para- meters that could assist in modeling suitability, proximity to high-voltage powers lines was chosen for site scoring because it provides a means for the power generated by biomass facilities to be put into the grid. This data set represents general locations of 115-, 161-, 230-, and 500-kilovolt power lines in the United States developed by the Federal Emergency Management Agency (FEMA). Data sets on the general location of high-voltage power line are available from NREL (2005).

These are just a few of the data sets that could be used in identifying locations for biomass projects in Georgia. Additional data sets of interest include road data, community tapestry data, species data, and land ownership/cadastral data, to name a few. Consultation with domain specialists is always recommended when modeling biomass site potential or other natural phenomena, to define essential variables, contributing factors to phenomena, and model parameters.

2)Analysis

Combining data and the method used to do so make analysis interesting. The answers we derive from the analysis make it useful. To determine site potential for biomass projects in Georgia, a simple additive approach was taken that ranked all locations based on summation of the respective parts (in this case the scoring of each aforementioned data set). The following paragraphs document the procedure.

The data sets not natively in raster format (specifically power lines and biomass potential at the county level) were first converted to raster in order to return a scoring at all locations in the study area (continuous across the state). The FEMA high-voltage power line data were buffered before conversion to raster using a multi-ringed buffer to provide varying proximities to power lines with which to score potential sites.

Once all of the data sets were converted, recoding of the data was performed. This was essential for two reasons: (1) Since NLCD data (land cover class data) are inherently nominal, mathematical computations on it were not possible; and (2) for the additive approach used, a common, "real" zero value was needed to rank specific locations as better or worse than other locations. The recoding procedure consisted of grouping values and then, in this case, giving the grouped values a score (a recoded value) that corresponded to their rank. Each input data set was recoded into 10 classes and scored sequentially from 1 to 10. Recoding required research and consultation with domain specialists to ensure that proper consideration and values were given to input variables.

After all the data sets were recoded, coincident raster cell values were summed to produce the final site potential score. Once summed, a data set scoring all locations in Georgia as to site potential for biomass projects ranging from 2 to 40 was generated. Once the data were responsible and cartographically represented, it could then be used by decision makers and developers to assist with biomass facility placement.

New Words and Phrase

1.disregard　*vt.*　不理会，忽视
2.lean　*vt.*　倚；依赖；使斜靠
3.amass　*vt.*　积累，积聚
4.conceptualization　*n.*　概念化
5.discrete　*adj.*　分离的
6.plethora　*n.*　过多，过剩
7.tabular　*adj.*　表格的
8.link　*vt.*　连接
9.explicit　*adj.*　明确的，清楚的
10.thematic　*adj.*　主题的
11.demarcate　*vt.*　定……的界线，区分
12.raster　*n.*　光栅
13.satellite　*n.*　卫星
14.spectrum　*n.*　光谱；波谱
15.undeniable　*adj.*　不可否认的
16.detriment　*n.*　损害，伤害
17.harvest　*n.*　结果
18.acquisition　*n.*　获得；购置
19.proximity　*n.*　接近，邻近

Notes

1.We abstract or simplify the real world into two types of spatial conceptualizations and corresponding models: the discrete view of objects in space (corresponding to vector representations) and the continuous view of those objects (analogous to raster representations).

我们将真实的世界抽象或者简化为两个空间概念和相应的模型：空间的离散视图（根据矢量表征）和连续视图（类似于光栅表征）。

analogous to　类似于（与 similar to 用法相同）

Traumatic purpura occurring in colloid milium may be analogous to that occurring in systemic amyloidosis.

2.However, energy derived from fossil fuels is undeniably finite and has been shown to be harmful to the environment during harvesting, manufacturing, and use.

尽管如此，化石燃料而来的能源毋庸置疑是有限的，并且在其采集、生成和使用过程中都会给环境带来危害。

derived from　来源于

Major part of the power is derived from the difference in pressure acting on front and back of runner blades and only a minor part from the dynamic action of velocity.

3.Once the data were responsible and cartographically represented, it could then be used by decision makers and developers to assist with biomass facility placement.

一旦数据通过地图形式表征，决策者或者开发者将采用它在生物通量设施布置中提供一定的参考。

once　一旦　assist with　在……方面给予帮助

Partnerships between public and private organizations aim to assist with everything from securing financing to writing a business plan.

Comprehensive Exercises

I.Answer the following questions on the text.

1.How do we abstract or simplify the real world?

2.Why did the debate in the GIS realm have virtually disappeared?

3.Why do GIS and data management will play a key role in shaping our world today and in the future?

4.What are the critical data sets in the discussions with foresters and research into biomass?

5.What can we learn from the case study?

II.Fill the most appropriate words or phrases in the correct forms in the blanks from the list below.

discrete	be useful for	representation	link	drop off
sustainable	available	perform	in terms of	consultation with

1.The availability of water in adequate quantity and quality is a necessary condition for________development.

2.In ancient times, bridge often crossed rivers at right angles, had humps in the middle, had no________footpaths, and were invariably built of stone and timber using the arch form of construction.

3.China has stayed in close communication and________not only the US, but other parties including Russia as well.

4.This capability______database object creation, design, and modification because it permits you to have a random order of CREATE commands.

5.In many urban areas transit buses use freeways to offer riders the fastest________trip.

6.A group of machines collectively called a concrete train _______ these operations at a rate of about three-fourths of a meter per minute.

7. _______ learning or other school I would come with strict demands on themselves, I will grasp this opportunity.

8.Vector________are appropriate for data sets having an explicit spatial location.

9.One of the ______in the water cycle is the important economic link wherein water is used to meet man's needs.

10.The average annual rainfall is over 1500mm, and _______gradually from south to north.

III.Translate the following sentences into Chinese from the text.

1.Vector models use points, lines, and polygons to respectively represent objects such as trees, transmission lines, and parcels of land. These geometric objects can contain a plethora of associated characteristics that in the vector model are traditionally stored as records in tables.

2.So far, we have not achieved a sustainable balance. GIS and sound data management are important components of sustainability. Whether they are used to justify socioeconomic decisions, determine the feasibility of renewable energy sources, or assess environmental degradation, GIS and data management will play a key role in shaping our world today and in the future. The following sections discuss such applications.

3.Additional data sets of interest include road data, community tapestry data, species data, and land ownership/cadastral data, to name a few. Consultation with domain specialists is always recommended when modeling biomass site potential or other natural phenomena, to define essential variables, contributing factors to phenomena, and model parameters.

4.Remote sensing is the science and art of gathering data from a location physically removed from the Earth's surface and then analyzing and scrutinizing that data. It is an often misunderstood technology victimized by over-ambitious expectations.

水利工程专业基础术语英文词汇集锦（1）

1.advanced maintenance dredging 备淤深度
2.adverse effects of blasting 爆破有害效应
3.aids to navigation 助航标志
4.aids to navigation on inland waterway 内河航标
5.allowable over-depth 允许超深
6.allowable over-width 允许超宽
7.dredger 挖泥船
8.antifouling panel 防污屏
9.approach channel 进港航道
10.approach channel 引航道
11.articulated Concrete Blocks 铰链排
12.artificial channel 人工航道
13.artificial fish nest 人工鱼巢
14.backflow 回流
15.bankfull discharge 平滩流量
16.bank protection works 护岸
17.bankwise mark 沿岸标
18.bar 拦门沙
19.bar stabilization 护滩
20.bar stabilization belt 护滩带
21.basin lock 广室船闸
22.bathymetric line 等深线
23.bathymetric survey 水深测量
24.bed load 推移质
25.bed material 河床质
26.bed material 底质
27.bed material load 床沙质

28.bedrock rapids 基岩急滩
29.bend-rushing flow 扫弯水
30.blow-off dredger 吹泥船
31.boil 泡水
32.bottom protection 护底
33.bottom protection belt 护底带
34.brook outlet rapids 溪口急滩
35.bucket dredger 链斗挖泥船
36.bulkhead gate 平板闸门
37.canalization works 渠化工程
38.canalized channel 渠化航道
39.capacity of navigation lock 船闸通过能力
40.central island 江心洲
41.chamber floor 闸底
42.channel chart 航道图
43.channel classes 航道等级
44.channel depth 航道水深
45.channel dimension 航道尺度
46.channel facilities 航道设施
47.channel improvement 航道治理
48.channel regulation 航道整治
49.channel width 航道宽度
50.chute cutoff 撇弯切滩
51.coastline 海岸线
52.cobble 卵石
53.containment bund 吹填围埝
54.cross section survey 横断面测量
55.culvert 输水廊道
56.cut-off works 裁弯工程
57.damming;diking 筑坝
58.dead water level 死水位
59.deep pool 深槽
60.deeps-staggered shoal 交错浅滩
61.delay blasting 延时爆破
62.density of freight traffic 货运密度
63.density of ship flow 船流密度
64.designed draft of typical ship 船舶设计吃水
65.designed lowest navigable stage 设计最低通航水位
66.designed minimum navigable discharge 设计最小通航流量
67.discharge hydrograph 流量过程线
68.dispersed filling and emptying system 分散输水系统
69.distorted model 变态模型
70.dredge-cut alignment 挖槽定线
71.dredge-cut design 挖槽设计
72.dredge-cut setting out 挖槽放线
73.dredger 挖泥船
74.ecological beach protection 生态护滩
75.ecological flexible mattress 生态排
76.effective dimensions of lock 船闸有效尺度
77.electric ignition 电力起爆
78.electronic detonator 电子雷管
79.electronic navigational chart 电子航道图
80.emulsion explosive 乳化炸药
81.enclosure 吹填围埝
82.end of backwater 回水末端
83.engineering geological investigation 工程地质调查
84.engineering geological surveying and mapping 工程地质测绘
85.entrance area 口门区
86.entrance channel 进港航道
87.estuarial channel 潮汐河口航道
88.estuarine river-flow reach 河口河流段
89.estuary 河口湾
90.estuary delta 河口三角洲
91.faintly-curved reach 微弯河段
92.filling and emptying valve 输水阀门

93.fissure water　裂隙水
94.fixed-bed river model　定床河工模型
95.bottom protection　软体排
96.floating mark　浮标
97.flood control water level　防洪限制水位
98.flood mark　泛滥标
99.flood observation　汛期观测
100.flood plain　河漫滩
101.flood stage　洪水位
102.gabion dam　石笼坝体
103.generalized physical model　概化模型
104.geographical range　地理视距
105.geological prospecting　地质勘探
106.geotextile sand container　沙被式软体排
107.geotextiles-laying vessel　铺排船
108.grab dredger　抓斗挖泥船
109.grain size distribution curve　颗粒级配曲线
110.gravel　砾石
111.lock' s navigable time　船闸通航时间保证率
112.guaranteed rate of navigation　通航保证率
113.horizontal tolerance　超宽
114.hybrid bank revetment　混合式护岸
115.hydraulic fill　吹填
116.intermittent reclamation　间歇吹填
117.kinds of aids layout　航标配布类别
118.load factor of locked ship　过闸船舶装载系数
119.lockage water　船闸用水量
120.lock canal　设闸运河
121.lock chamber　闸室
122.magmatic rock　岩浆岩
123.magnetic prospecting　磁力测探
124.maintenance dredging　维护性疏浚
125.maintenance gate　检修闸门
126.maintenance of waterway　航道维护
127.management of dredged material　疏浚土管理
128.marine abrasion rock　海蚀岩
129.marking depth　设标水深
130.mean velocity in section　断面平均流速
131.mean velocity on a vertical　垂线平均流速
132.median stage　中水位
133.metamorphic rock　变质岩
134.mid-channel slope measuremen　河心比降测量
135.minimum burden　最小抵抗线
136.misfire　盲炮
137.miter gate　人字闸门
138.mobilization of dredger　挖泥船调遣
139.modal shift over the junction　翻坝转运
140.model calibration　模型率定
141.model material　模型沙
142.model sand　模型沙
143.model scale　模型比尺
144.model test　模型试验
145.movable-bed model with suspended load　悬移质动床模型
146.movable-bed river model　动床模型
147.multiple locks　多线船闸
148.Multistep locks　多级船闸
149.national waterway　国家航道
150.natural channel　天然航道
151.navigable clear height　通航净高
152.navigable clear width　通航净宽
153.navigable dimension　通航尺度
154.navigable river　通航河流

155.navigable stage control of canal 运河通航水位控制
156.navigable stretch 通航河段
157.navigation hydrojunction 航运枢纽
158.navigation-hydropower junction 航电枢纽
159.navigation light 航标灯
160.navigation lock 船闸
161.navigation mark 航行标志
162.normal pool level 正常蓄水位
163.open navigation canal 开敞式运河
164.pebble rapids 卵石急滩
165.pebble shoal 卵石浅滩
166.pool design flood level 水库设计洪水位
167.quality index of rock 岩石质量指标
168.quality of dredge-cut 挖槽质量
169.rapids 急滩
170.reclamation 吹填
171.regulation stage 整治水位
172.regulation width 整治线宽度
173.reservoir 水库
174.reservoir clearance 清库
175.river engineering model 河工模型
176.river pattern 河型
177.river terrace 河流阶地
178.river valley 河谷
179.sand 沙
180.sand and pebble shoal 砂卵石浅滩
181.sand shoal 沙质浅滩
182.scissors-like flow 剪刀水
183.sea canal 通海运河
184.sea chart 海图
185.sediment discharge 输沙率
186.settlement volume 沉降量
187.ship chamber space 承船厢室
188.ship incline 斜面升船机
189.ship lift 升船机
190.ship lock 船闸
191.ship model test 船模航行试验
192.shoal 浅滩
193.shoal and rapids 滩险
194.side flat 边滩
195.side scan sonar inspection 侧扫声呐检测
196.siltation 回淤
197.special purpose waterway 专用航道
198.spit 沙嘴
199.spoil-dyke 吹填围埝
200.standard curvature radius of channel 航道标准弯曲半径
201.standard depth of channel 航道标准水深
202.standard dimensions of waterway maintenance 航道维护标准尺度
203.submerged bar 潜洲
204.submerged closure dike 潜锁坝
205.submerged groin 潜丁坝
206.submerged longitudinal dike 潜顺坝
207.suspended load 悬移质
208.undistorted model 正态模型
209.valley flat 河漫滩
210.water slope shiplift 水坡式升船机
211.water stage 水位
212.waterway milestone 航道里程
213.waterway survey 航道测量
214.whistling mark 鸣笛标
215.wide chamber lock 广室船闸
216.wind direction 风向
217.wind-induced wave 风成波

UNIT V CONSTRUCTION ENGINEERING

Lesson 1 Building Materials

The term "building materials" refers to materials used in the construction trade, and which are generally classified in categories based on their type; Metals (e. g. steel, aluminum, copper), minerals (natural stone, concrete, glass) and organic materials (e. g. wood, plastic, bitumen). Modern building materials cannot always be readily placed in one of these groups, as they may be developed through the systematic combination of different types to form composites that provide improved properties over the individual materials themselves (composite building materials).A major factor in the development of building materials is that new structures are being asked to perform increasingly multifaceted tasks. In addition to their traditional load-bearing capacities and use as room partitions, building materials also need to fulfil a multitude of additional functions today. Along with technical criteria, economic and ecological criteria have become increasingly important factors when choosing and developing building materials. Materials with the smallest possible environmental impact (such as low levels of toxic emissions or required primary energy) are considered sustainable and suitable for use in the future.

In the planning phase for buildings and construction, the most important aspects include technical features (e. g. mechanical stability, deformation capacity, and fire protection) and the materials' physical and structural characteristics.

The technology of building materials has developed into a true science; studies are performed on the structural characteristics of materials using research methods common to both materials science and solid state physics. Based on this information, specific building material characteristics can be optimized for specific applications by performing targeted structural alterations (down to the nano scale). Building materials such as these are often referred to as "tailor-made building materials".

For a variety of reasons, concrete is the most frequently used building material in the world. The required raw materials are usually readily available and inexpensive. Due to its rapidly obtained plastic properties after mixing, concrete can be used for an almost unlimited range of designs. The finished product (in its hardened state) is highly stable and offers good long-term durability.

Concrete principally consists of five raw materials: an aggregate filler, cement (which reacts hydraulically with water, the other main ingredient), and the so-called additives and admixtures, which improve the physical and chemical properties of fresh and hardened concrete.

The additives are mostly reactive or inert powder materials, such as pulverized stone or industrial by products (pulverized fly ash, granulatedblast furnace slag). Admixtures are usually liquids developed through chemical construction processes (polymer-based plasticisers are one example). Cements are binding agents which solidify both in air and under water, and which harden to become stone-like and water-resistant. Compared to other hydraulic binding agents such as hydraulic lime, cements offer much higher stability and strength. Lime and clay or their naturally occurring compound (marl lime) are mainly used as raw materials for manufacturing cement. The hardening process of cement bound materials after mixing with water is known as hydration. Due to complex gel and crystal formation processes, aqueous reaction products are formed in the cement clinker stage, or hydrate phases. In the early phase of hydration, gel formation is predominant with respect to crystal formation. Gel formation causes the cement paste to set. The weak bonds between the dispersed particles can be broken however by mechanical disturbances, resulting in gel re-liquefaction. This phenomena, known as thixotropic behaviour, is used to produce concrete construction elements with improved compaction (vibration compaction). Vibrators are employed to oscillate the concrete and obtain mechanical compaction.

1.Self-compacting concrete

Through the use of high-capacity plasticisers, the viscosity of the cement-bound materials can be altered so that no exogenous compacting work is required to obtain optimized compaction (self-compacting concrete, SCC).

SCC has a honey-like consistency and compacts under its own weight. SCC applications may be used for dense, reinforced construction components; for complicated, geometrical construction components; as well as for manufacturing exposed concrete surfaces that are almost entirely nonporous.

2.Ultra-high performance concrete

By reducing the water content and the diameter of the coarse aggregate grain, and by using reactive (e. g. pulverised fly ash, silica fume) and nonreactive (e. g. stone powder) additives, the packing density of a cement-bound material can be increased to its maximum level. The maximum particle diameter of the aggregate in ordinary concrete is lowered from 16-32 mm to between 1-0.063 mm for ultra-high performance concrete (UHPC). The interstices between the aggregates and the cement-stone matrix are filled with the above- mentioned ingredients. In addition to the physical filling, the by-products which accumulate in the cement hydration (such as portlandite-$Ca(OH)_2$ are transformed into mineral phases (calcium silicate hydrate phases-CSH-phases) which increase strength and durability. The water needs the dry mixture to increase in reverse proportion to the particle size (aggregate, stone powder). In order to ensure good workability, liquefying polymer-based admixtures are added.

By reducing the recognized weak points in the texture (such as micro cracks, shrinkage cracks and cavity pores), the maximum possible load-bearing capacity of the building material is increased. What's more, the higher density increases resistance to chemical and physical exposure. Brittleness, which increases along with the compressive strength, is reduced by adding suitable types of fibres. These improved materials with increased durability can help reduce the load-bearing cross-section of a structural member, making it possible to build constructions which up to now were unimaginable for bridges, high-rise buildings or factories.

3.Composite building materials

Composites consist of two or more materials which are positively linked. By combining several different material characteristics, the advantages of the individual materials can be used, while the disadvantages are discarded. Several systems exist, depending on the type of composite:

a)particle reinforced composites.

b) laminated composites.

c) fibres reinforced composites.

4. Textile concrete

For the manufacture of textile concretes, textile structures made of high- performance fibres are used. For reinforced concrete structures, a minimum coverage is required between the steel reinforcement and the surfaces which are exposed or otherwise stressed due to risks of corrosion. Since there is no risk of corrosion with alkaliresistant textile fibres, they are particularly well suited for filigree and thinwalled construction components, and are now used for manufacturing façade components and, more recently, for pedestrian bridges and shell structures as well.

5.Fibre concrete

The primary use of fibres is to improve the marginal tensile strength of solid building materials and the associated risks of cracking. Cement-bound building materials can only withstand low tensile forces, and their brittle material behaviour is problematic when the maximum load is exceeded. But, by combining materials with a high tensile strength, composite materials with improved material qualities can be manufactured. Different types of fibre materials, quantities and geometries are added to the concrete depending on the field of application. Due to their high tensile strength, steel fibres are often used. Since steel fibres tend to be vulnerable to corrosion, alternative materials such as stainless steel, glass, plastic and carbon fibres are used in certain applications.

Concrete behaviour in fires and the risk of shrinkage cracks can be improved by adding plastic fibres (polypropylene, polyethylene). In a fire, the temperature can rise to over 1000 ℃ within a few minutes. When concrete heats at such a rapid rate, the pore water in the concrete component evaporates.

Spalling is caused by the formation of a quasi-saturated layer as a result of water vapour condensation flowing in from colder zones. The lower the permeability of the concrete, the faster these gastight layers occur, and the higher the resulting vapour pressures. The water barrier which is formed ("moisture clog") is impermeable to any further water vapor. If the water vapour pressure exceeds the tensile strength of the concrete, spalling will occur.

The positive effect of PP-fibres on the high-temperature behaviour of concrete is based on the improved permeability before and during exposure to fire. Additional micropores are formed when fibres are added. The resulting transition zones between the fibres and the cementstone matrix consist of additional pores and weak cement hydration products (ettringite, portlandite). The transition zones and the contact surfaces of the aggregate form a permeable system even before exposure to high temperatures. As temperature increases, the PP-fibres melt and burn up, creating more capillary pores in the hardened concrete. In this way, the entire system obtains higher filtration capacity and allows the increasing vapour pressure in the interior to escape.

6.Wood-Polymer-Compound

WPC is a three-material system, which mainly consists of wood and thermoplastic materials, and a smaller proportion of additives. For economic reasons, pine wood in various forms and quantities is added to the plastic matrix to manufacture WPC. The wood-powder filler content is between 50% and 75%. Since the natural wood material is damaged when the temperature exceeds 200 ℃ and the mechanical qualities of the final product are irretrievably degraded, plastics must be used which have a melting and processing temperature under 200℃ . For this reason, conventional mass-produced plastics such as polyethylene (PE), polyvinyl chloride (PVC) or polypropylene (PP) are increasingly being used for WPC. Additives markedly change the material properties of the processed material and the end product.

In order to modify processing qualities, conventional additives such as plasticisers and lubricants are used. Plasticisers increase the elasticity of the plastics, and lubricants decrease the interior and exterior friction of the plastic melt. The material qualities of the final product are altered by introducing additives such as stabilisers, dyes, fillers and reinforcing materials. Flame-retarding additives are also used. Adhesion promoter is the decisive additive for the interaction of polar wood fibres and non-polar plastics. The adhesion promoter improves the bond, which improves the mechanical properties. At the same time, water needs the wood particles to decrease because they have been completely surrounded by polymer matrix.

Before carrying out the different manufacturing techniques, the raw materials are dried and mixed to form a homogeneous mixture. As an alternative to energy intensive drying processes, water binding materials such as starch can also be used. For today's applications, WPC is mainly used for linear profiles using extrusion technology. In extrusion technology, the solid mass of the raw materials is heated to 200 ℃ to form a homogeneous, plasticised melt, and is then pressed out by an extrusion tool at 100-300 bar. Various units are located in front of and behind the extrusion line, which perform cooling, finishing, shaping, cutting to size, and curing.

New Words and Phrase

1.aluminium *n.* 铝
2.copper *adj.* 铜
3.concrete *n.* 混凝土
4.bitumen *n.* 仪器沥青
5.mulitifaceted *adj.* 多方面的
6.fulfil *n.* 支流实现
7.multitude 大量解释，说明
8.toxic *n.* 尺度有毒的
9.tailor *n.* 定做变形，失真
10.raw *vi.* 未经加工的 (使) 倾斜
11.durability *n.* 耐久性
12.aggregate *n.* 集料
13.cement *n.* 水泥多边形 ; 多角形
14.additive *adj.* 添加剂
15.inert *adj.* 惰性的
16.pulverize *n.* 粉碎
17.binding agent 黏合剂
18.lime *n.* 石灰
19.gel *n.* 凝胶
20.crystal *n.* 结晶 ; 晶体
21.hydration *n.* 水化
22.vibrator *n.* 震动器
23.exogenous *adj.* 外源性的
24.ash *n.* 灰烬
25.interstice *n.* 裂缝 ; 空隙
26.brittleness *n.* 脆性
27.tensile *adj.* 张力的
28.textile *n.* 纺织品 ; 纺织业
29.fibre *n.* 纤维
30.moisture *n.* 含水率
31.thermoplastic *adj.* 热塑
32.pine *n.* 松树
33.irretrievably *adj.* 不能挽回地
34.bridge *vi.* 跨过，度过
35.starch *n.* 淀粉
36.extrusion *n.* 挤压

Notes

1.The term “building materials” refers to materials used in the construction trade, and which are generally classified in categories based on their type; metals (e. g. steel, aluminium, copper), minerals (natural stone, concrete, glass) and organic materials (e. g. wood, plastic, bitumen).

建筑材料是指在建造工程中所应用的材料，其主要基于他们的类型被分为：金属（铁、铝、铜），矿物（天然石头、混凝土、玻璃）以及有机材料（比如木材、塑料以及沥青等）。

based on 以……为基础，基于

Most structural elements in a berth structure can be prefabricated and the choice of precast elements in the superstructure will mainly be based on economic considerations.

2.Cements are binding agents which solidify both in air and under water, and which harden to become stone-like and water-resistant.

水泥是将空气和水固化的黏合剂，逐渐变得像石头一样坚硬且防水。

harden 坚固，硬化 water-resistant 防水

Mould the mixture into shape while hot, before it hardens.

3.By reducing the water content and the diameter of the coarse aggregate grain, and by using reactive (e. g. pulverised fly ash, silica fume) and nonreactive (e. g. stone powder) additives, the packing density of a cement-bound material can be increased to its maximum level.

通过减少含水率和集料混合物中的粒径，以及采用可反应的（粉煤灰、硅灰）和不可反应的（岩粉）添加剂，水泥材料的压实密度可增加至其最大极限值。

by 通过，后面可接动名词 packing density 压实密度

By reducing the recognized weak points in the texture, the maximum possible load-bearing capacity of the building material is increased.

Comprehensive Exercises

I.Answer the following questions on the text.

1.What are the types of building materials?

2.What dose the concrete consists of?

3.What should we consider in the planning phase for construction?

4.How to improve the packing density of the cement-bound material?

5.What includes in the extrusion line?

II.Fill the most appropriate words or phrases in the correct forms in the blanks from the list below.

based on	optimise	aggregate	reactive	oscillate
unimaginable	filigree	spall	pull out	in front of

1.Each ‘perfume’ of each investment in product development, the need to pay attention, time and money are _____.

2.Now it will increasingly be used to automate and _____interactions with the physical environment.

3.It involves measuring a network of triangles that are _____points on the earth’s surface.

4.This jewelry shows some of the earliest examples of metalworking methods such as_____and granulation.

5.There are things you _____ the picture because people can relate to them.

6.To prevent _____, the manufacturer must develop spall resistant rolls while the user must be familiar with the roll characteristics and maintain the rolls in a manner to promote longevity.

7.This is a very exposed point because of the danger of slides _____the rear wall or retaining friction slab.

8.There have always been slight ______ in world temperature.

9.China is ranked the sixth in the world in terms of ______annual water resources.

10.The additives______or inert powder materials, such as pulverized stone or industrial by products.

III.Translate the following sentences into Chinese from the text.

1.The term "building materials" refers to materials used in the construction trade, and which are generally classified in categories based on their type; metals (e. g. steel, aluminium, copper), minerals (natural stone, concrete, glass) and organic materials (e. g. wood, plastic, bitumen).

2.By reducing the water content and the diameter of the coarse aggregate grain, and by using reactive (e. g. pulverised fly ash, silica fume) and nonreactive (e. g. stone powder) additives, the packing density of a cement-bound material can be increased to its maximum level.

3.In order to modify processing qualities, conventional additives such as plasticisers and lubricants are used. Plasticisers increase the elasticity of the plastics, and lubricants decrease the interior and exterior friction of the plastic melt. The material qualities of the final product are altered by introducing additives such as stabilisers, dyes, fillers and reinforcing materials.

4.As a result, cracks develop whenever loads, or restrained shrinkage or temperature changes, give rise to tensile stresses in excess of the tensile strength of the concrete. In a plain concrete beam, the moments about the neutral axis due to applied loads are resisted by an internal tension-compression couple involving tension in the concrete.

Lesson 2　Surveying

Before any civil engineering project can be designed, a survey of the site must be made. Surveying means measuring and recording by means of maps—the earth's surface with the greatest degree of accuracy possible. Some engineering projects—highways, dams, or tunnels, for example—may require extensive surveying in order to determine the best and most economical location or route.

Measurements in a survey include distance, elevations (heights of feature within the area), boundaries (both man-made and natural), and other physical characteristics of the site. Some of these measurements will be in a horizontal plane; that is, perpendicular to the force of gravity. Others will be in a vertical plane, in line with the direction of gravity. The measurement of angles in either the horizontal or vertical plane is an important aspect of surveying in order to determine precise boundaries or precise elevations.

There are two kinds of surveying: plane and geodetic. Plane surveying is the measurement of the earth's surface without curvature. Within areas of about 20 kilometers square—the earth's curvature does not produce any significant errors in a plane survey. For large areas, however, a geodetic survey, which takes into account the curvature of the earth, must be made.

1. Plane surveying

In plane surveying, the principal measuring device for distance is the steel tape. In English-speaking countries, it has replaced a rule called a chain, which was either 66 or 100 feet long. The 66-feet-long chain gave speakers of English the acre, measuring ten square chains or 43560 square fees as a measure of land area. The men who hold the steel tape during a survey are still usually called chainmen. They generally level the tape by means of plumb bobs, which are lead weights attached to a line that give the direction of gravity. When especially accurate results are required, other means of support, such as a tripod—a stand with three legs—can be used. The indicated length of a steel tape is in fact exactly accurate only at a temperature of 200 centigrade, so temperature readings are often taken during a survey to correct distances by allowing for expansion or contraction of the tape.

Distance between elevations is measured in a horizontal plane. In the diagram alongside, the distance between the two hills is measured from point *A* to *B* rather than from points *A* to *C* to *D* to *B*. When distances are being measured on a slope, a procedure called breaking chain is followed. This means that measurements are taken with less than the full length of the tape（Fig.5.1）.

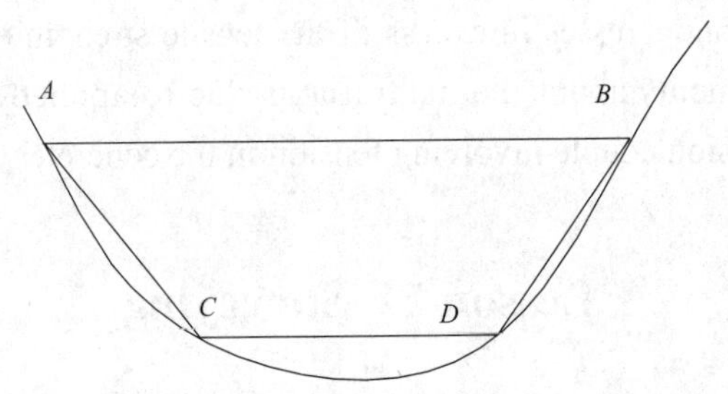

Figure 5.1 Distance between two hills

Lining up the tape in a straight line of sight is the responsibility of the transitman, who is equipped with a telescopic instrument called a transit. The transit has plates that can indicate both vertical and horizontal angles, as well as leveling devices that keep it in a horizontal plane. Cross hairs within the telescope permit the transitman to line up the ends of the tape when he has them in focus.

Angles are measured in degrees of arc. Two different systems are in use. One is the sexagesimal system that employs 3600, each degree consisting of 60 minutes and each minute of 60 seconds. The other is the centesimal system that employs 400 grads, each grad consisting of 100 minutes and each minute of 100 seconds. A special telescopic instrument that gives more accurate readings of angles that the transit is called a theodolite.

In addition to cross hairs, transits and theodolites have markings called stadia hairs (stadia are the plural of the Greek word stadion, a measure of distance). The stadia hairs are parallel to the horizontal cross hair. The transitman sights a rod, which is a rule with spaces marked at regular intervals. The stadia hairs are fixed to represent a distance that is usual a hundred times each of the marks on the rod. That is, when the stadia hairs are in line with a mark on the rod that reads 2.5, the transit is 250 meters from the rod. Stadia surveys are particularly useful in determining contour lines, the lines on a map that enclose areas of equal elevation.

Contour maps can be made in the field by means of a plane-table alidade. The alidade is a telescope with vertical circle and stadia hairs. It is mounted on a straight-edged metal plane that can be kept parallel to the line of sight. The surveyor can mark his readings of distances and elevations on a plane (or flat) table that serves as a drawing board. When the marks representing equal elevations are connected, the surveyor has made a contour map.

Heights or elevations are determined by means of a surveyor's level, another kind of telescope with a bubble leveling device parallel to the telescope. A bubble level, which is similar to a carpenter's level, is a tube containing a fluid that has an air bubble in it. When the bubble is centered in the middle of the tube, the device is level. The surveyor sights a rule called a level rod through the telescope. The rod is marked off to show units of measure in large, clear number. The spaces between the marks usually are alternately black and white in order to increase visibility. The number that the surveyor reads on the level rod, less the height of his or her instrument, is the vertical elevation.

Heights are given in relation to other heights. On maps for example, the usual procedure is to give the elevation above sea level. Sea level, incidentally, can be determined only after averaging the tides in a given area over a definite period. A survey carried out by level and rod often gives the elevation in relation to a previously measured point that is called a bench mark.

Approximate elevations can also be measured with an altimeter, which is a device that takes advantage of changes in atmospheric pressure. Readings taken with an altimeter are usually made at two, and sometimes three, different points and then averaged. The readings must be corrected for humidity and temperature, as well as the weight of air itself.

Modern technology has been used for surveying in instruments that measure distance by means of light or sound waves. These devices direct the waves toward a target that reflects them back to a receiver at the point of origin. The length of time it takes the waves to go to the target and return can then be computed into distance. This surveying method is particularly useful when taking measurements over bodies of water.

Aerial photography is another modern method of surveying. A photograph distorts scale at its edges in proportion to the distance the subject is from being in a direct vertical line with lens of camera. For this reason, the photographs for an aerial survey are arranged to overlap so that the scale of one part joins the scale of the next. This arrangement is called a mosaic, after the pictures that are made from hundreds of bits of colored stone or grass.

2.Geodetic surveying

Geodetic surveying is much more complex than plane surveying. It involves measuring a network for triangles that are based on points on the earth's surface. The triangulation is then reconciled by mathematical calculations with the shape of the earth. This shape, incidentally, is not a perfect sphere but an imaginary surface, slightly flattened at the poles, that represents mean sea level as though it were continued even under the continental land masses.

In addition to measuring surfaces for civil engineering projects, it is often necessary to make a geological survey. This involves determining the composition of the soil and rock that underlie the surface at the construction site. The nature of the soil, the depth at which bedrock is located, and the existence of faults or underground streams are subsurface factors that help civil engineers determine the type and size of the structural foundations or the weight of the structure that can rest on them. In some areas, these can be critical factors. For example, Mexico City rests on a lakebed with no bedrock near the surface; it is also located in an earthquake zone. The height and weight of buildings must therefore be carefully calculated so that they will not exceed the limits that are imposed by the site.

Geological samples are most often obtained by borings, in which hollow drills bring up cores consisting of the different layers of underground materials. Other devices that are used in geological survey are gravimeters and magnetometers. The gravimeter measures the earth's gravitational pull; heavier rocks like granite exert a stronger pull than lighter ones like limestone. The magnetometer measures the strength of the earth's magnetic field. Again, the denser the rock, the more magnetic force it exerts.

A third instrument is the seismograph, which measures vibrations, or seismic waves, within the earth. It is the same instrument that is used to detect and record earthquakes. In a geological survey, setting off small, man-made earthquakes uses it. The waves created by a blast of dynamite buried in the ground reflect the different kinds of rock under the surface; hard or dense rocks reflect the waves more strongly than soft or porous rocks.

New Words and Phrase

1.geodetic *adj.* 大地测量学的
2.perpendicular *adj.* 垂直的
3.chain *n.* 链条
4.acre *n.* 恶化，堕落
5.tape *n.* 卷尺
6.plumb bob 铅锤
7.tripod *n.* 三脚架
8.telescopic *adj.* 望远镜的
9.transit *n.* 运输，搬运
10. sexagesimal system 六十分进制
11. centesimal system 百分制
12. theodolite *n.* 经纬仪
13. cross hair 十字线
14. rod *n.* 杆
15. contour line *n.* 等高线
16. alidade *n.* 照准仪，测高仪
17. bubble *n.* 气泡
18. humidity *n.* 湿度，热度
19. Aerial photography 空中摄影
20. overlap *n.* 重叠
21. mosaic *n.* 马赛克
22. reconcile *v.* 使和谐一致，调和
23. imaginary *adj.* 幻想的
24. gravimeter *n.* 重力仪

25. magnetometer　*n.*　磁力计
26. seismograph　*n.*　地震仪
27. vibration　*adj.*　震动
28. porous　*adj.*　多孔的，透水的

Notes

1.The measurement of angles in either the horizontal or vertical plane is an important aspect of surveying in order to determine precise boundaries or precise elevations.

与平面或者垂直面的夹角测量是现场测量的重要方面，其决定了边界或者高程的准确性。

either or　或……或……，表示两者之一，连接句子中两个并列的成分。

Precipitation is either absorbed into the ground or runs off into rivers.

in order to　为了，以便于

Business and industrial streets may require up to six lanes in order to handle local traffic and access needs.

2.Heights or elevations are determined by means of a surveyor's level, another kind of telescope with a bubble leveling device parallel to the telescope.

测量人员的水平决定了测量的高度或者高程，另外一种望远镜是具有气泡调平装置的望远镜。

by means of　用，依靠　parallel to　平行于

By means of powerful hydraulic jacks, the span, weighing 5200 tons, was to be raised from barges to a height of 46 m.

3.Geological samples are most often obtained by borings, in which hollow drills bring up cores consisting of the different layers of underground materials.

测量样本通常通过地勘获取，采用空心钻提取地下不同层的组成物质。

boring　开凿，钻　consist of　组成

During the work of boring, all the excavated material must be brought out through the narrow confines of the bore itself.

Comprehensive Exercises

I.Answer the following questions on the text.

1.What is the difference between plane surveying and geodetic surveying?

2.What are transits and theodolites?

3.What are triangulation? Is it used in plane or geodetic surveying?

4.What do a magnetometer measure?

5.What type of channel pattern forms when the inflowing sediment discharge exceeds the outgoing sediment capacity?

II.Fill the most appropriate words or phrases in the correct forms in the blanks from the list below.

by means of	take into account	chain	lead to	transit
parallel to	alternately	in relation to	rest on	porous

1.A comparison of the measured and calculated incipient velocity from formulas widely used in the Yangtze River showed that Tang's formula ______the influence of water pressure and the molecular force.

2.It is also concerned with disposing of excess water and waste materials______sewer systems.

3.In the Warren truss, the diagonals______ are in tension and compression.

4.These are tied together by ______and carry a dismanteable platform over them, which forms the carriageway.

5.It is preferable to standardize the column cross-sections and rather vary the amount of reinforcement in the columns ______the column length and load.

6.Floods ______the damages of roads, collapse of bridges or traffic congestion, which may affect the daily operation.

7.The wings of a span in the movable bridge______the channel.

8.Usually the foundations for the piers ______bedrock, and often underwater.

9.Seepage is the slow leaking of water through ______material, such as earth or some kinds of rock like limestone or sandstone.

10.The electrical power, which is consequently obtained at the terminals of the generator, ______the area where it is to be used for doing work.

III.Translate the following sentences into Chinese from the text.

1.Surveying means measuring-and recording by means of maps—the earth's surface with the greatest degree of accuracy possible. Some engineering projects—highways, dams, or tunnels, for example—may require extensive surveying in order to determine the best and most economical location or route.

2.Contour maps can be made in the field by means of a plane-table alidade. The alidade is a telescope with vertical circle and stadia hairs. It is mounted on a straight-edged metal plane that can be kept parallel to the line of sight. The surveyor can mark his readings of distances and elevations on a plane (or flat) table that serves as a drawing board.

3.In addition to measuring surfaces for civil engineering projects, it is often necessary to make a geological survey. This involves determining the composition of the soil and rock that underlie the surface at the construction site. The nature of the soil, the depth at which bedrock is located, and the existence of faults or underground streams are subsurface factors that help civil engineers determine the type and size of the structural foundations or the weight of the structure that can rest on them.

4.Other devices that are used in geological survey are gravimeters and magnetometers. The gravimeter measures the earth's gravitational pull; heavier rocks like granite exert a stronger pull than lighter ones like limestone.

Lesson 3 Construction Engineering

Construction engineering is a specialized branch of civil engineering concerned with the planning, execution, and control of construction operations for such projects as highways, buildings, dams, airports, and utility lines.

Planning consists of scheduling the work to be done and selecting the most suitable construction methods and equipment for the project. Execution requires the timely mobilization of all drawings, layouts, and materials on the job to prevent delays to the work. Control consists of analyzing progress and cost to ensure that the project will be done on schedule and within the estimated cost.

1.Planning

The planning phase starts with a detailed study of construction plans and specifications. From this study a list of all items of work is prepared, and related items are then grouped together for listing on a master schedule. A sequence of construction and the time to be allotted for each item is then indicated. The method of operation and the equipment to be used for the individual work items are selected to satisfy the schedule and the character of the project at the lowest possible cost.

The amount of time allotted for a certain operation and the selection of methods of operation and equipment that is readily available to the contractor. After the master or general construction schedule has been drawn up, subsidiary detailed schedules or forecasts are prepared from the master schedule. These include individual schedules for procurement of material, equipment, and labor, as well as forecasts of cost and income.

2.Execution

The speedy execution of the project requires the ready supply of all materials, equipment, and labor when needed. The construction engineer is generally responsible for initiating the purchase of most construction materials and expediting their delivery to the project. Some materials, such as structural steel and mechanical equipment, require partial or complete fabrication by a supplier. For these fabricated materials the engineer must prepare or check all fabrication drawings for accuracy and case of assembly and often inspect the supplier's fabrication.

Other construction engineering duties are the layout of the work by surveying methods, the preparation of detail drawings to clarify the design engineer's drawings for the construction crews, and the inspection of the work to ensure that it complies with plans and specifications.

On most large projects it is necessary to design and prepare construction drawings for temporary construction facilities, such as drainage structures, access roads, office and storage buildings, formwork, and cofferdams. Other problems are the selection of electrical and mechanical equipment and the design of structural features for concrete material processing and mixing plants and for compressed air, water, and electrical distribution systems.

3.Control

Progress control is obtained by comparing actual performance on the work against the desired performance set up on the master or detailed schedules. Since delay on one feature of the project could easily affect the entire job, it is often necessary to add equipment or crews to speed up the work.

Cost control is obtained by comparing actual unit costs for individual work items against estimated or budgeted unit costs, which are set up at the beginning of the work. A unit cost is obtained by dividing the total cost of an operation by the number of units in that operation.

Typical units are cubic yards for excavation or concrete work and tons for structural steel. The actual unit cost for any item at any time is obtained by dividing the accumulated costs charged to that item by the accumulated units of work performed.

Individual work item costs are obtained by periodically distributing job costs, such as payroll and invoices to the various work item accounts. Payroll and equipment rental charges are distributed with the aid of time cards prepared by crew foremen. The cards indicate the time spent by the job crews and equipment on the different elements of the work. The allocation of material costs is based on the quantity of each type of material used for each specific item.

When the comparison of actual and estimated unit costs indicates an overrun; an analysis is made to pinpoint the cause. If the overrun is in equipment costs, it may be that the equipment has insufficient capacity or that it is not working properly. If the overrun is in labor costs, it may be that the crews have too many men, lack of proper supervision, or are being delayed for lack of materials or layout. In such cases time studies are invaluable in analyzing productivity.

4.Project Quality Management

Project Quality Management includes the processes required to ensure that the project will satisfy the needs for which it was undertaken It includes 'all activities of the overall management function that determine the quality policy, objectives, and responsibilities and implements them by means such as quality planning, quality control, quality assurance, and quality improvement, within the quality system.

1) Quality Planning-identifying which quality standards are relevant to the project and determining how to satisfy them.

2) Quality Assurance-evaluating overall project performance on a regular basis to provide confidence that the project will satisfy the relevant quality standards.

3) Quality Control-monitoring specific project results to determine if they comply with relevant quality standards and identifying ways to eliminate causes of unsatisfactory performance.

These processes interact with each other and with the processes in the other knowledge areas as well. Each process may involve effort from one or more individuals or groups of individuals based on the needs of the project. Each process generally occurs at least once in every project phase.

Quality is 'the totality of characteristics of an entity that bears on its ability to satisfy stated or implied needs.' A critical aspect of quality management in the project context is the necessity to turn implied needs into stated needs through project context is the necessity to turn implied needs into stated needs through project scope management.'

The project management team must be careful not to confuse quality with grade. Grade is 'a category or rank given to entities having the same functional use but different requirements for quality'. Low quality is always a problem; low grade may not be. For example, a software product may be of high quality (no obvious bugs, readable manual) and low grade (a limited number of features), or of low quality (many bugs, poorly organized user documentation) and high grade (numerous features). Determining and delivering the required levels of both quality and grade are the responsibilities of the project manager and the project management team.

The project management team should also be aware that modern quality management complements modern project management. For example, both disciplines recognize the importance of:

Customer satisfaction-understanding, managing, and influencing needs so that customer expectations are met or exceeded. This requires a combination of conformance to specifications (the project must produce what it said it would produce) and fitness for use (the product or service produced must satisfy real needs).

Prevention over inspection-the cost of avoiding mistakes is always much less than the cost of correcting them.

Management responsibility-success requires the participation of all members of the team, but it remains the responsibility of management to provide the resources needed to succeed.

Processes within phases-the repeated plan-do-check-act cycle described by Deming and others is highly similar to the combination of phases and processes discussed in Project Management Processes.

In addition, quality improvement initiatives undertaken by the performing organization (e.g., TQM, Continuous Improvement, and others) can improve the quality of the project management as well as the quality of the project product.

However, there is an important difference that the project management team must be acutely aware of the temporary nature of the project means that investments in product quality improvement, especially defect prevention and appraisal, must often be borne by the performing organization since the project may not last long enough to rap the rewards.

New Words and Phrases

1.execution *v.* 执行，完成，实施，施工
2.maste *adj.* 主要的，总的，熟练的
3.clarify *v.* 澄清，阐明，净化，解释
4.budget *n.* 预算；*v.* 做预算，编入预算
5.payroll *n.* 薪水册，发放工资额，工资单
6.invoice *n.* 发票，发货单，开发票，记清单
7.overrun *v.* 超过，超出，超限
8.schedule *n.* 进度表
9.undertake *vt.* 承担，担任，许诺，保证；*v.* 采取
10. implement *n.* 工具，器具；*vt.* 贯彻，实现；*v.* 执行
11. overlap *v.* （与……）交叠
12. overwork *n.* 过度操作，过度工作；*v.*（使）工作过度
13. entity *n.* 实体
14. confuse *vt.* 搞乱，使糊涂
15. deliver *vt.* 递送，陈述，释放，发表（一篇演说等），交付
16. exceed *vt.* 超越，胜过；vi. 超过其他
17. initiative *n.* 主动
18. interface *n.* （地质）分界面，接触面，（物、化）界面
19. identify *vt.* 识别，鉴别，把……和……看成一样；*v.* 确定
20. proprietary *adj.* 所有的，私人拥有的；*n.* 所有者，所有权
21. appraisal *n.* 评价，评估（尤指估价财产，以便征税），鉴定
22. project quality management 项目质量管理
23. overall management 全面管理
24. quality policy 质量方针
25. quality planning 质量计划
26. quality control 质量控制
27. quality assurance 质量保证
28. quality improvement 质量改进
29. quality system 质量体系
30. quality standard 质量标准
31. on a regular basis 定期地
32. comply with 照做
33. approach to quality management 质量管理方法
34. the international Organization for Standardization（ISO）国际标准组织
35. Total Quality Management(TQM) 全面质量管理
36. project stakeholder 项目利害关系者，项目利益相关者
37. project team 项目组
38. employee turnover 雇员更新
39. quality inspection 质量检查
40. project scope management 项目范围管理
41. project manager 项目经理
42. customer satisfaction 客户满意
43. customer expectation 顾客期望

Notes

1.Construction engineering is a specialized branch of civil engineering concerned with the planning, execution, and control of construction operations for such projects as highways, buildings, dams, airports, and utility lines.

施工工程是土木工程学科的一个专门的分支，其涉及高速公路、建筑、修坝、机场和基础设施的规划、执行以及施工过程的管理控制。

concerned with　涉及；说到；关于

Because engineering is concerned with actions to be taken in the future, an important part of the engineering process is improving the certainty of decisions with respect to satisfying the objectives of engineering applications.

2.Project Quality Management includes the processes required to ensure that the project will satisfy the needs for which it was undertaken.

项目质量管理包括项目满足其需求所需的过程。

此句中 for which it was undertaken 是定语从句，修饰 the needs；which 指代 the need, it 指代 the project。

3.It includes 'all activities of the overall management function that determine the quality policy, objectives, and responsibilities and implements them by means such as quality planning, quality control, quality assurance, and quality improvement, within the quality system'.

它包括"确定质量方针、目标和职责的整个管理职能的全部活动，并在质量体系中通过诸如质量计划、质量控制、质量保证和质量改进等方法来实施"。

此句中 that determine the quality policy，objectives，and responsibilities 是定语从句，修饰 all activities of the overall management function；them 指代 all activities of the overall management function。

Grade is 'a category or rank given to entities having the same functional use but different requirements for quality'.

Comprehensive Exercises

I.Answer the following questions on the text.

1.What is the construction engineering?

2.What is the construction engineer responsible for?

3.What is the Project Quality Management?

4.How do we improve the quality of the engineering project?

5.What is the project management team must be acutely aware of?

II.Fill the most appropriate words or phrases in the correct forms in the blanks from the list below.

concerned with	schedule	responsible for	obtain	invaluable
eliminate	imply	interact with	confuse	aware of

1.We are______negative side-effects of engineering innovations (such as air pollution from automobiles) than ever before.

2.This does not______ the lack of progress in the study of this world water balance element.

3.Geological samples ______by borings, in which hollow drills bring up cores consisting of the different layers of underground material.

4.Photographs provide ______ records of changing conditions.

5.The project management team must be careful not to_______quality with grade.

6.As a result of this complexion reaction, soap cannot_______ the dirt on clothing, and the calcium-soap complex itself forms undesirable precipitates.

7.Storage is provided to meet peak demands and to allow the plant to operate on a uniform______.

8.Reducing liquid water losses or controlling their runoff is mostly achieved with terraces and contour planting to reduce or ______long steep slopes.

9.This paper _______ the method to simulate the sediment transport process.

10.This is owing in part to the fact that we do not fully understand the natural processes _______ the movement of the various contaminants.

III.Translate the following sentences into Chinese from the text.

1.Planning consists of scheduling the work to be done and selecting the most suitable construction methods and equipment for the project. Execution requires the timely mobilization of all drawings, layouts, and materials on the job to prevent delays to the work.Individual work item costs are obtained by periodically distributing job costs, such as payroll and invoices to the various work item accounts. Payroll and equipment rental charges are distributed with the aid of time cards prepared by crew foremen.

2.A critical aspect of quality management in the project context is the necessity to turn implied needs into stated needs through project context is the necessity to turn implied needs into stated needs through project scope management.

3.Rolling concrete dams, as their name says, are made up of rolling concrete(RC). It is a material which due to the small quantity of binding material and water is more like, 'wet earth', than like massive concrete. It contains the same basic components as with the conventional concrete: aggregate, binding medium, water and additives (in case of need).

Lesson 4 Engineering Economy

Ages ago, the most significant barriers to engineers were technological. The things that engineers wanted to do, they simply did not yet know how to do, or hadn't yet developed the tools to do. There are certainly many more challenges like this which face present-day engineers. However, we have reached the point in engineering where it is no longer possible, in most cases, simply to design and build things for the sake simply of designing and building them. Natural resources (from which we must build things) are becoming scarcer and more expensive. We are much more aware of negative side-effects of engineering innovations (such as air pollution from automobiles) than ever before.

1. Why do engineers need to learn about economics?

For these reasons, engineers are tasked more and more to place their project ideas within the larger framework of the environment within a specific planet, country, or region. Engineers must ask themselves if a particular project will offer some net benefit to the people who will be affected by the project, after considering its inherent benefits, plus any negative side-effects (externalities), plus the cost of consuming natural resources, both in the price that must be paid for them and the realization that once they are used for that project, they will no longer be available for any other projects.

Simply put, engineers must decide if the benefits of a project exceed its costs, and must make this comparison in a unified framework. The framework within which to make this comparison is the field of engineering economics, which strives to answer exactly these questions, and perhaps more. The Accreditation Board for Engineering and Technology (ABET) states that engineering 'is the profession in which a knowledge of the mathematical and natural sciences gained by study, experience, and practice is applied with judgment to develop ways to utilize, economically, the materials and forces of nature for the benefit of mankind'.

It should be clear from this discussion that consideration of economic factors is as important as regard for the physical laws and science that determine what can be accomplished with engineering. Satisfaction of the physical and economic environments is linked through production and construction processes. Engineers need to manipulate systems to achieve a balance in attributes in both the physical and economic environments, and within the bounds of limited resources. Following are some examples where engineering economy plays a crucial role:

1) Choosing the best design for a high-efficiency gas furnace.

2) Selecting the most suitable robot for a welding operation on an automotive assembly line.

3) Making a recommendation about whether jet airplanes for an overnight delivery service should be purchased or leased.

4) Considering the choice between reusable and disposable bottles for high-demand

beverages.

With items 1 and 2 in particular, note that coursework in engineering should provide sufficient means to determine a good design for a furnace, or a suitable robot for an assembly line, but it is the economic evaluation that allows the further definition of a best design or the most suitable robot.

There are numerous examples of engineering systems that have physical design but little economic worth (i.e it may simply be too expensive!!). Consider a proposal to purify all of the water used by a large city by boiling it and collecting it again through condensation. This type of experiment is done in junior physical science labs every day, but at the scale required by a large city, is simply too costly.

2.Role of uncertainty in engineering

When conducting engineering economic analyses, it will be assumed at first, for simplicity, that benefits, costs, and physical quantities will be known with a high degree of confidence. This degree of confidence is sometimes called assumed certainty. In virtually all situations, however, there is some doubt as to the ultimate values of various quantities. Both risk and uncertainty in decision-making activities are caused by a lack of precise knowledge regarding future conditions, technological developments, synergies among funded projects, etc. Decisions under risk are decisions in which the analyst models the decision problem in terms of assumed possible future outcomes, or scenarios, whose probabilities of occurrence can be estimated. Of course, this type of analysis requires an understanding of the field of probability. Decisions under uncertainty, by contrast, are decision problems characterized by several unknown futures for which probabilities of occurrence cannot be estimated. Other less objective means exist for the analysis of such problems.

For the purposes of this brief tutorial, we cannot delve further into the analytical extensions required to accommodate risk or uncertainty in the decision process. We must recognize that these things exist, however, and be careful about reaching strong conclusions based on data which might be susceptible to these. Because engineering is concerned with actions to be taken in the future, an important part of the engineering process is improving the certainty of decisions with respect to satisfying the objectives of engineering applications.

3. The engineering process

Engineering activities dealing with elements of the physical environment take place to meet human needs that arise in an economic setting. The engineering process employed from the time a particular need is recognized until it is satisfied may be divided into a number of phases:

1) Determination of objectives. This step involves finding out what people need and want that can be supplied by engineering. People's wants may arise from logical considerations, emotional drives, or a combination of the two.

2) Identification of strategic factors. The factors that stand in the way of attaining objectives are known as limiting factors. Once the limiting factors have been identified, they are examined to locate strategic factors - those factors which can be altered to remove limitations restricting the success of an undertaking. A woman who wants to empty the water from her swimming pool might be faced with the limiting factor that she only has a bucket to do the job with, and this would require far greater time and physical exertion than she has at her disposal. A strategic factor developed in response to this limitation would be the procurement of some sort of pumping device which could do the job much more quickly, with almost no physical effort on the part of the woman.

3) Determination of means (engineering proposals). This step involves discovering what means exist to alter strategic factors in order to overcome limiting factors. In the previous example, one means was to buy (or rent) a pump. Of course, if the woman had a garden hose, she might have been able to siphon the water out of the pump. In other engineering applications, it may be necessary to fabricate the means to solve problems from scratch.

4) Evaluation of engineering proposals. It is usually possible to accomplish the same result with a variety of means. Once these means have been described fully, in the form of project proposals, economic analysis can be employed to determine which among them, if any, is the best means for solving the problem at hand.

5) Assistance in decision making. It is commonplace for the final decision-making responsibility to fall on the head of someone other than the engineers. The person so charged, however, may not be sufficiently knowledgeable about the technical aspects of a proposal to determine its relevant worth compared to other means. The engineer can help to bridge this gap.

4. Engineering economic studies

The four key steps in planning an economic study are:

1) Creative step: People with vision and initiative adopt the premise that better opportunities exist than are known to them. This leads to research, exploration, and investigation of potential opportunities.

2) Definition step: System alternatives are synthesised with economic requirements and physical requirements, and enumerated with respect to inputs/outputs.

3) Conversion step: The attributes of system alternatives are converted to a common measure so that systems can be compared. Future cash flows are assigned to each alternative, consisting of the time-value of money.

4) Decision step: Qualitative and quantitative inputs and outputs to/from each system form the basis for system comparison and decision making. Decisions among system alternatives should be made on the basis of their differences. For a small number of real world systems there will be complete knowledge. All facts/information and their relationships, judgements and predictive behavior become a certainty. For most systems, however, even after all of

the data that can be bought to bear on it has been considered, some areas of uncertainty are likely to remain. If a decision must be made, these areas of uncertainty must be bridged by consideration of non-quantitative data/information, such as common sense, judgement and so forth.

New Words and Phrase

1.innovation *n.* 创造；创新
2.externality *n.* 溢出效应
3.realization *n.* 认识；领会
4.strive *vi.* 努力；奋斗
5.crucial *adj.* 至关重要，关键性的
6.furnace *n.* 熔炉
7.robot *n.* 同步化机器人
8.analyst *n.* 分析者
9.delve *vi.* 翻找
10. synthesis *n.* 综合；结合

Notes

1.However, we have reached the point in engineering where it is no longer possible, in most cases, simply to design and build things for the sake simply of designing and building them.

尽管如此，现在在很多情况下我们已经很难仅仅通过规划和设计的目的出发来进行设计和规划工程。

该句型中 in most cases 为状语，翻译过程中通常可放在句尾。

One of the reasons they have survived is because of the great strength that was built into themstrength greater than necessary in most cases.

2.Once the limiting factors have been identified, they are examined to locate strategic factors-those factors which can be altered to remove limitations restricting the success of an undertaking.

一旦找到了限制因素，他们将通过策略因子的检查——这些因子可以用于消除限制因素带来的实施工程过程中的限制。

strategic factors 策略因子 undertake 承担；从事；负责

These and other problems might have been prevented by more thorough studies before construction was undertaken.

3.It is commonplace for the final decision-making responsibility to fall on the head of someone other than the engineers.

最后工程的决定权一般在其他人头上而不是工程师。

commonplace 普遍的 other than 除了……以外

If all fluxes other than evapotranspiration can be assessed, the evapotranspiration can be deduced from the change in soil water content over the time period.

Comprehensive Exercises

I.Answer the following questions on the text.

1.Why do engineers need to learn about economics?

2.How do the engineers consider the balance between project and cost?

3.What are the role of uncertainty in engineering?

4.What do the engineering process consist of?

5.What are the key steps in planning an economic study?

II.Fill the most appropriate words or phrases in the correct forms in the blanks from the list below.

aware of	strive	utilize	with respect to	attain
in response to	convert	alternative	bear	consideration

1.People ______have pleasant smelling homes and offices, and industry, for many years, has supplied various kinds of air freshening products to oblige them.

2.Heavy metals, organic pollutants, pathogenic organisms, etc., should all be understood ______their viability in natural systems and ability to cause environmental damage.

3.The study must ______not only structural features, but also economic factors and possible alternatives, or other choices.

4.This will be a major accomplishment, and doubts that the means of_______it were not always the most efficient should not detract from it.

5.The load-bearing capacity of columns of 80 cm diameter and more is seldom fully _____, and the lengths of beam and slab spans are determined rather by the dimensions of the beams and the slabs themselves.

6.The most important of these is probably a vibrating roller, which compacts the earth until it ______the weight of the base course and wearing surface that will rest on it.

7.Progressive abutment deformation or yielding _______arch thrust results in load-transfer and stress redistribution within the dam shell and in the abutment itself.

8.The appropriate factor to_____an investment _____an equivalent annual cost is designated as the capital recovery factor and may be computed from the expression where represents the interest rate per annum and represents the years of estimated life.

9.The usual _______has been concrete columns of minimum 70 cm diameter founded as described above for pillar quays.

10.It also includes _______of possible alternatives, such as a concrete gravity dam or an earth-fill embankment dam.

III.Translate the following sentences into Chinese from the text.

1.Simply put, engineers must decide if the benefits of a project exceed its costs, and must make this comparison in a unified framework. The framework within which to make this comparison is the field of engineering economics, which strives to answer exactly these questions, and perhaps more.

2.It is commonplace for the final decision-making responsibility to fall on the head of someone other than the engineers. The person so charged, however, may not be sufficiently knowledgeable about the technical aspects of a proposal to determine its relevant worth compared to other means. The engineer can help to bridge this gap.

3.It is commonplace for the final decision-making responsibility to fall on the head of someone other than the engineers. The person so charged, however, may not be sufficiently knowledgeable about the technical aspects of a proposal to determine its relevant worth compared to other means. The engineer can help to bridge this gap.

4.Generally speaking, each choice among alternatives should be made on economic grounds. Each alternative that is given serious consideration should be expressed in money units, before the choice is made. In fact, unless alternatives can be expressed in money units, the items involved in such choices are incommensurable.

水利工程专业基础术语英文词汇集锦（2）

1.sea port　海港

2.estuary port　河口港

3.reservoir port　水库港

4.design vessel　设计船型

5.draft　吃水

6.full loaded draft　满载吃水

7.light draft　空载吃水

8.ship motions　船舶运动

9.port planning　港口规划

10. general layout of port　港口总体布置

11. port hinter land　港口腹地

12. port operating district　港口作业区

13. goods traffic　货运量

14. freight turnover　货运周转量

15. cargo throughput of port　港口货物吞吐量

16. handling technology　装卸工艺

17. volume of cargo　泊位作业量

18. container crane　集装箱起重机

19. approach channel　进港航道

20. anchorage area　锚地

21. water depth in front of wharf　码头前水深

22. bulk cargo terminal　散货码头

23. specialized terminal　专业化码头

24. roll-on roll-off terminal　滚装码头

25. berth　泊位

26. port water supply system　港口给水

27. berth capacity　泊位通过能力

28. port design capacity　港口设计能力

29. utility factor of berth　泊位利用率

30. wharf　顺岸码头

31. finger pier　突堤码头

32. offshore terminal　离岸码头

33. approach trestle pier　引桥式码头

34. open sea terminal　开敞式码头

35. quay wall　实体式码头

36. quay wall　直立式码头

37. sloping wharf　斜坡式码头

38. gravity quay wall　重力式码头

39. pile foundation　桩基
40. flexible platform　柔性承台
41. filter layer　倒滤层
42. settlement joint　沉降缝
43. approach trestle　引桥
44. crane track　码头起重机轨道
45. breakwater　防波堤
46. hydraulic reclamation　吹填
47. geotextile　土木织物
48. packet sand drains　袋砂井
49. caisson launching　沉箱下水
50. piling base line　沉桩基线
51. dredger　挖泥船
52. floating crane　起重船
53. boring ship　钻探船
54. surveying ship　测量船
55. crane track　码头起重机轨道
56. dock road　港口道路
57. dumping and filling on land　陆上抛填
58. entrance channel　金刚航道
59. multi-purpose terminal　多用途码头
60. project quality management　项目质量管理
61. overall management　全面管理
62. quality policy　质量方针
63. quality planning　质量计划
64. quality control　质量控制
65. quality assurance　质量保证
66. quality improvement　质量改进
67. quality system　质量体系
68. quality standard　质量标准
69. approach to quality management　质量管理方法
70. the International Organization for Standardization（ISO）国际标准组织
71. Total Quality Management(TQM)　全面质量管理
72. project stakeholder 项目利害关系者，项目利益相关者
73. project team　项目组
74. employee turnover　雇员更新
75. quality inspection　质量检查
76. project scope management　项目范围管理
77. project manager　项目经理
78. customer satisfaction　客户满意
79. customer expectation　顾客期望
80. performing organization　执行组织
81. construction general layout　施工总平面布置
82. construction programming　施工组织
83. aerated spillway　真空式溢流道
84. barrage-type spillway　堰式溢洪道
85. berm spillway　护道排水沟
86. chute spillway　陡槽式溢洪道
87. drop spillway　跌水式溢洪道
88. lateral flow spillway　测流式溢洪道
89. morning glory spillway　喇叭形直井溢洪道
90. overflow spillway　溢洪道
91. shaft spillway　直井式溢洪道
92. siphon spillway 虹吸溢水道

UNIT VI CHALLENGE FOR HYDRAULIC ENGINNERING

Lesson 1 Climate Change in River System

Climate affects freshwater ecosystems indirectly through societal and economic systems, as well as directly by temperature and precipitation. This context focuses on how climate–hydromorphological interactions might alter freshwater ecosystems. We include projected changes and their effects at the catchment, reach and habitat scales. The emphasis is on streams and rivers, although some aspects of lake hydromorphology are also considered. Finally, we discuss how climate change might affect attempts to restore stream and river ecosystems.

1.Effects of climate change at the catchment scale

1.1 Direct climate impacts on stream hydrology

The effects that changes in climate may have on stream hydrology are illustrated by the expected changes in discharge for the River Lambourn in southern England (Fig. 6.1). The Lambourn catchment (265 km^2) has a limited river network owing to its underlying porous chalk geology and is dominated by arable land and improved pasture. Based on RCAO HadAM3H model and a catchment-scale rainfall-stream flow model, the mean monthly discharge for 2071-2100 was predicted. The current and modelled discharge time series were compared for magnitude, frequency and timing of extreme discharge events. A more variable discharge regime for the river Lambourn by the end of this century is predicted. The biggest difference between now and then is the size of the winter peaks. During 1974-95, 45% of annual peaks exceeded 3 m^3s^{-1}, while for 2071-2100, 83% of peaks are predicted to exceed this level. There were two prolonged drought events in 1974-95, caused by a lack of sufficient winter recharge in 1976 and again in 1991 and 1992. The modelled discharge data for 2071-2100 predict that equivalent winter droughts are less likely to occur. Even the poorest winter recharge levels for the projected period are comparable with current normal recharge levels. These predicted changes to discharge magnitude, frequency and timing are likely to affect the physical structure of the River Lambourn and its riparian zone, as well as the flora and fauna it supports.

Climate change will thus alter the dominant pattern of precipitation, which in turn changes run-off and discharge regimes, including spates and droughts in streams and rivers. Spates and droughts are important drivers that contribute to stream patchiness. Climate change, therefore, presents a challenge to the structure and function of current stream ecosystems. During spates, habitats may be destroyed; during low flows, they will be silted, and during base-flow conditions, habitats will be generated again.

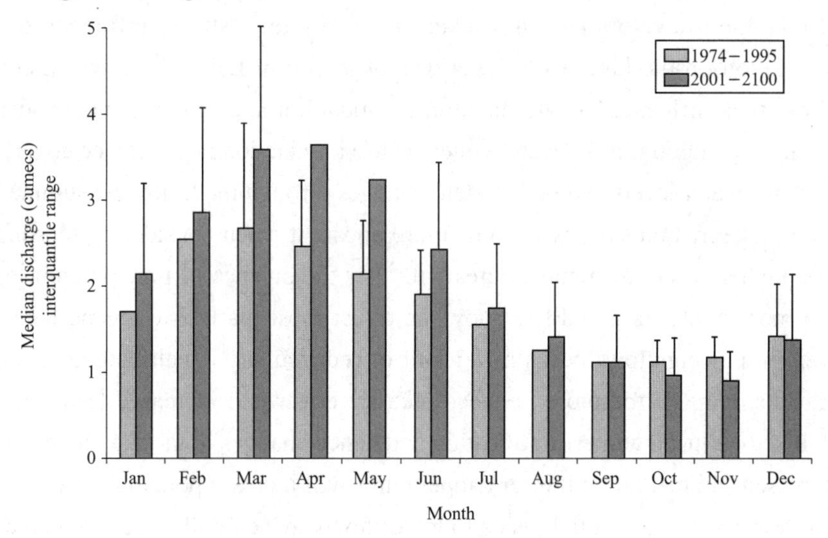

Figure 6.1 Current and projected discharge of the river Lambourn, southern England

The predictability of resources will decrease, and species will have to adapt or become locally extinct. High flows and spates scour accumulated sediment and debris, redistribute streambed material and organic matter in the channel, change channel morphology and form new erosion (runs and riffles) and deposition (point and mid-channel bars, pools, sand accumulations) zones. High flows and spates may also disturb in-channel and encroaching riparian vegetation, homo genize water chemistry among the stream channel and adjacent water bodies and increase shear stress on organisms. In contrast, low flows and droughts bring siltation of fine mineral and organic material, decrease oxygen concentrations and increase those of some nutrients and minerals.

Change in flow variability and timing can also be important, resulting in unpredictable erosion-deposition processes with frequent shifts in channel morphology and habitat availability and loss of synchronization of flow stages with stages in organisms'life cycles, such as egg deposition, growth and pupation.

1.2 Direct climate change impacts on lake hydrology

Climate change will impact lake hydrology mainly through effects on residence time and water level as well as through receptors and sources of stream flow. Short residence times mean that pollutants such as excess nutrients from point sources are flushed out of the lake ecosystem, whereas with decreasing precipitation and longer residence times, they will accumulate, with likely changes in phytoplankton communities and in food-web composition and structure. In lakes with long residence times, internal processes may become more important. For example, phytoplankton production may increase with higher temperatures due to increased nutrient availability, and eutrophication problems may thereby become more severe.

A decline in water level due to decreased precipitation may cause changes in the nutrient status and acidity of lakes with low buffering capacities. Water-level change can also directly affect phytoplankton development in a lake. For example, a strong influence of the North Atlantic Oscillation on the lake water level was observed in Lake Vortsjarv (Estonia) where water-level changes influenced phytoplankton composition and biomass independently of the nutrient loading. In addition, less severe winters cause a reduction in winter ice cover, which can lead to lower lake water levels and lake system changes through the following summer months.

Effects on lake hydrology may become more apparent when considering seasonal patterns. For example, higher winter air temperatures will alter the balance of precipitation input so that there will be more winter rain and less snow, an effect most marked in upland lakes. This will lead to lower spring 'highflow' peaks as a result of reduced snow melt and, as a consequence, less flushing in spring. Furthermore, reduced lake ice cover and increased frequency of winter storms will lead to reduced winter stratification and hence changes in in-lake biogeochemistry. In contrast, increased summer air and water temperatures will increase potential evaporation, which, coupled with lower summer rainfall, leads to lower flows, which will affect both lake levels and downstream river communities. Increasing episodic events such as storms, at whatever season they occur, will lead to greater flushing and run-off. More frequent or more severe episodes will affect lake and outflow chemistry such that, for example, acid episodes may be more frequent. Greater winter run-off will also lead to a higher suspended sediment load, which will lead to increases in lake sediment accumulation rate (SAR).

2.The effects of climate change on hydraulics and morphology at the reach scale

The pattern and variation of current velocity within a stream reach has a major influence on longitudinal and transverse channel morphology, species diversity, food-web structure and ecological processes. It is often assumed that monthly or daily means, which are often readily available, are sufficient to characterize flow regime. However, even single events can cause substantial changes in the physical habitat and can affect ecological functioning. Five major hydraulic conditions affect the distribution and ecological success of stream biota: suspended load, bed-load movement, water-column effects, such as turbulence and velocity profile, and substratum interactions (near-bed hydraulics). However, stream organisms are generally adapted to a wide variability in stream discharge and can accommodate large changes.

A study carried out in the River Lambourn assessed how hydraulic changes at the reach scale affected benthic macroinvertebrate community composition. The results suggested considerable resilience in stream macroinvertebrate communities to future changes in discharge regimes. Changes in the cover of the five main mesohabitats (Berula, Ranunculus and Callitriche plant stands, gravel and silt) were recorded over 9 years. The macroinvertebrate community associated with each mesohabitat was also sampled. Berula and Ranunculusstands were more abundant in early summers preceded by high- discharge winters, while gravel habitats tended to be more prominent after winters with lower discharge. The plant and mineral mesohabitats proved to be distinct from each other, with the caddis fly families, Glossosomatidae and Goeridae, being relatively more abundant in gravel, the bivalves, Sphaeriidae, favouring silt patches and the mayfly families, Ephemerellidae and Caenidae, preferring the plants. However, taxa did not strongly associate with just one particular mesohabitat; they tended to be equally abundant on two or more mesohabitats. Perhaps because of the lack of strong obligate relationships between taxa and specific mesohabitats, the early summer density of most taxa was not correlated with previous winter discharge levels. But distribution across a range of available habitats means that the macroinvertebrate community is buffered against yearly variations in discharge regime, as long as there are no extreme spates resulting in strong currents. These can cause catastrophic downstream drift due to increased shear stress, especially for small animals.

The increase in discharge extremes (both spates and droughts) and the decrease in discharge predictability expected with changing climate may increase the relative importance of abiotic parameters, both hydraulic conditions and physical habitat features, in shaping future stream communities. Spates have substantial effects. Some organisms may move or be moved to habitats with less disturbance (e.g. stream margins, the hyporheic zone or other patches protected from high flow) and, thus, will seek refuges. Others may be able to remain in or on relatively stable habitats where they are sheltered from shear stress. Many may lack adaptations and be washed downstream.

When disturbances become highly unpredictable, species might display riskspreading strategies, for example, asynchronous hatching of eggs. Behavioural adaptations enable organisms to respond directly to individual floods and droughts and unstable substrates by moving to refuges, following use of cues that they can detect. Droughts and low flows elicit other strategies. There may be physiological resistance, adapted life cycles, allowing early maturation and flight of adults, or resting stages. There may also be increased mobility with organisms seeking out remaining higher-flow habitats. Climate change may also result in death through sediment injury or burial, increased drift downstream, physiological problems for respiration, as particles settle on gill filaments, and through low oxygen concentrations in the water due to the deposition of fine organic sediments. There may also be food shortages as food sources (organic matter, periphyton) are buried by settling particles or scoured by the current. There can be an increase of the sediment concentration in suspension in the water column (affecting filter feeders positively) or a decrease in potential prey (loss of individuals due to drift, burial) for predators.

3.The effects of climate change on stream and river restoration success

After a long period of modification of streams and rivers, as well as their catchments, to the needs of agriculture, industry and households, there has been an increasing awareness of the large negative impact of these alterations. In the Netherlands, only about 4% of the streams still have a natural morphology and a (more or less) natural hydrology. In Denmark, only 2% are more or less natural, and in Germany, this is 2%-5%. Environmental awareness and concern for the loss of stream and floodplain habitats and associated biodiversity have stimulated a major programme of stream rehabilitation and restoration, especially physical stream restoration in Europe. For example, in the Netherlands, 70 projects were carried out in 1991, 170 in 1993 and over 200 in 1998 with a total cost by 2006 estimated at 1.3 billion euro.

There are many ways of physically restoring streams such as reforestation of the floodplain, re-meandering and the removal of dams and bank structures. Newer approaches include the addition of coarse woody debris, the removal of sediment deposits in floodplains and various methods to combat the deep cutting of streams.

For effective stream restoration, the complex links among physical parameters, habitat diversity and biodiversity need to be understood. When a stream has been physically restored, success, measured mainly by an increase in biodiversity, depends on the extent of re-colonization by the original (indicator) species. This is the 'field of dreams' hypothesis, which states 'if we build it, they will come'. However, this is by no means a guaranteed outcome. Whether the original species are able to re-colonize the restored stream depends not only on the quality of the restored habitat but also on a number of factors such as the dispersal capacities of the species and the presence or absence of migration barriers between the source populations and the restored areas.Establishment of an invasive or non-native species may also hinder re-colonization, and biodiversity may, in general, be threatened by invasive species replacing native ones. Colonization may be less successful than hoped for, as illustrated by a study from the German Central Highlands, where the benthic macroinvertebrate biodiversity of multiple channels re-created in seven rivers was compared with that of seven engineered straight channels.

The hydromorphological diversity of the multiple-channel sections approached the intended reference condition. Habitat diversity increased, and increased sediment dynamics were observed. However, substrata in single- and multiple-channel sections still had similar macroinvertebrate community composition, and the alpha diversity of substratum-specific communities did not change. Different substrata, however, did host distinct macroinvertebrate communities, so beta diversity was greater in the restored channels as they tended to have a greater range of substrata. A comparison of representative communities in single- and multiple-channel sections showed very high Bray–Curtis similarities (69%-77%). Mean similarity analyses (using MEANSIM) revealed that the macroinvertebrate community composition within a channel type was less similar (single-channel within-group similarity: 0.61, multiplechannel within-group similarity: 0.63) than the composition between channel types (between-group similarity: 0.66).Differences between paired stream sections could mainly be attributed to single taxa that occurred solely in either the single- or multiple-channel sections. These exclusive taxa were mostly found on organic substrates such as living parts of terrestrial plants, large wood, coarse particulate organic matter and mud. Whether this occurrence is related to the specific substrata or to chance is still to be resolved. The overall high similarity of macroinvertebrate communities from single- or multiple-channel sections could be due to prevailing large-scale catchment pressures, the small scale of the restoration or a lack of potential re-colonizers. In general, river restoration schemes have proved similarly disappointing for these reasons.One of the most important keys to restoration is understanding the complex links between physical, chemical and biological components, ultimately at the catchment scale. Stream restoration is usually focussed on individual stream reaches, largely ignoring the importance of connectivity, such as channel movement for terrestrial–aquatic linkages (e.g. stream-floodplain exchanges) and biotic dispersal (Verdonschot & Nijboer 2002). Catchment-scale processes largely determine the structure and function of streams and their floodplains.

Thus, the scale of restoration needs to be adjusted to the scale of the dominant shaping processes. The key themes in stream ecology deal with the four dimensions of hierarchy of physical organization, adaptation, response of species and human disturbance. A catchment approach includes all these themes. Climate change, and thus hydromorphological change, affects the stream at the highest hierarchical levels, implying that under changing climate conditions, either the measures to meet restoration targets must be increased or the targets must be less ambitious.

4.Conclusions

Climate change will alter the hydromorphological conditions of lakes, streams and rivers. The magnitude of change induced by climate is still relatively small in comparison with the impact of anthropogenic land use, but in future, climate change may cause significant change in hydrology and, at the same time, impose land-use changes in catchments. Changes in stream hydrology are best characterized in terms of dynamics, particularly associated with increases in drought as well as in spate frequency. In lakes, hydrological changes are expressed in terms of more dynamic fluctuations as well as overall changes in water level and their impact on eutrophication . In rivers, climate change may also increase flow variability resulting in higher scouring and siltation rates, and where rivers flow into lakes, sediment loads may increase, leading to accelerated SARs. To some extent, biota in streams are adapted to changes in habitat-scale conditions and will tolerate such dynamics.

However, different life stages require different environmental conditions, and a disturbed timing, for example, of snow melt and the connected high-flow conditions, can result in a loss of taxa. More dynamic conditions and a loss of native taxa widen the opportunity for nonnative species to enter ecosystems. Globalized transport systems and hydrological links between large catchments further enhance this process.Restoration of streams and rivers is promoted by the EU Water Framework Directive and other legislations. Restoration success, however, has been lower than expected. This failure is mainly due to the scale of the restoration being too small in many cases, a lack of an ecosystem approach and a lack of understanding of key biological processes governing dispersal and colonization rates and the role of barriers. Future climate change is likely to reduce further the chances of restoration success.

New Words and Phrase

1.pasture *n.* 牧场；牧草地
2.flora *n.* 植物群
3.spate *n.* 泛滥；洪水
4.patchiness *n.* 小片 斑块
5.extinct *adj.* 灭绝的
6.encroach *v.* 侵占；侵犯
7.synchronization *n.* 同步化
8.pupation *n.* [昆]化蛹
9.receptor *n.* 感受器；接受器
10. phytoplankton *n.* 浮游生物
11. eutrophication *n.* 富营养化
12. buffer *v.* 缓冲；减轻
13. episodic *adj.* 插曲式的；偶然发生的；短暂的
14. asynchronous *adj.* 异步的
15. elicit *vt.* 引出；诱出
16. periphyton *n.* [生]固着生物；水生附着生物
17. burial *n.* 埋葬；葬礼
18. colonization *n.* 殖民；殖民地化
19. hierarchy *n.* 等级制度
20. invasive *adj.* 侵入的

Notes

1.Climate change will thus alter the dominant pattern of precipitation, which in turn changes run-off and discharge regimes, including spates and droughts in streams and rivers.

气候变化将引起降雨主要模式的改变，继而引起径流和流量的改变，包括河流洪水和干旱的改变。

alter （使）改变，更改，改动；修改　　in turn 继而；转而；反过来；逐一；依次；轮流地

Around the outer surface of the caisson was wrapped tarred felt, and this in turn was covered with planking.

2.The pattern and variation of current velocity within a stream reach has a major influence on longitudinal and transverse channel morphology, species diversity, food-web structure and ecological processes.

河道内流速的分布和变化将主要影响河道纵向和横向的地貌、物种种类、食物网结构以及生态过程。

influence 影响；可为动词或者名称，动词后常接 on/over 等

Businesses make large contributions to members of Congress, hoping to influence their votes on key issues.

3.In rivers, climate change may also increase flow variability resulting in higher scouring and siltation rates, and where rivers flow into lakes, sediment loads may increase, leading to accelerated SARs.

气候变化可能使得河道内流速增加，引起冲刷和淤积速率的增加，从而在汇入河流后带来更多的泥沙，引起泥沙淤积的累积。

result in 引起，导致，后面通常接结果

The fall in the value of the yen might result in a fractional increase in interest rates of perhaps a quarter of one percent.

Comprehensive Exercises

I.Answer the following questions on the text.

1.What are the direct climate impacts on stream hydrology?

2.What can the change of the current velocity in a stream lead to?

3.How many types of hydraulic condition affect the stream biota?

4.What is the effective stream restoration?

5.What will take place with the climate change in the world?

II.Fill the most appropriate words or phrases in the correct forms in the blanks from the list below.

focus on	illustrate	spate	contribute to	apparent
runoff	siltation	buffer	variability	Restoration

1.Most of the sources result from _____and leaks or seepage of pollutants into surface water or groundwater.

2.A compromise has to be reached between all the powerful vested interests before any ______work in the city can take place.

3.This plan requires the cooperation of government and_____watersheds and nonpoint sources of pollution, particularly runoff from urban areas and agricultural land.

4.Channels in bays and ocean waterways, as in the intra-coastal waterway also subjected to continued ______of varying magnitude.

5.Figures ______some of the concepts of esthetic design.

6.During _______, habitats may be destroyed; during low flows, they will be silted, and during base-flow conditions, habitats will be generated again.

7.The world's natural ecosystems act as______to create a suitable environment for human.

8.The proliferation of products, such as the wide variety o paper goods now produced in a range of colors,______the amount and complexity of waste discharges.

9._______the two leaves failed to latch together properly. But the bridge has been repaired and is now functioning successfully.

10.Longer-lived trees evolve more slowly and show less ______ in their rates of evolution.

III.Translate the following sentences into Chinese from the text.

1.Climate affects freshwater ecosystems indirectly through societal and economic systems, such as land management, as well as directly by temperature and precipitation. In many cases, climate change is an additional stressor adding to the impacts of human activity. Freshwater biodiversity, for example, is at present affected by the over-exploitation of natural resources, water pollution, flow modification, habitat degradation and by invasive alien species.

2.In contrast, increased summer air and water temperatures will increase potential evaporation, which, coupled with lower summer rainfall, leads to lower flows, which will affect both lake levels and downstream river communities. Increasing episodic events such as storms, at whatever season they occur, will lead to greater flushing and run-off.

3.Changes in stream hydrology are best characterized in terms of dynamics, particularly associated with increases in drought as well as in spate frequency. In lakes, hydrological changes are expressed in terms of more dynamic fluctuations as well as overall changes in water level and their impact on eutrophication.

Lesson 2 Environmental Consideration in Inland Navigation

With the passage and implementation of the national environmental Policy Act (NEPA) of 1969, environmental impact assessments of water resource projects under the US Army Corps of Engineers and other Federal agencies assumed a greater level of importance. Previously, environmental assessments were controlled by internal regulations and were usually not distributed or reviewed outside the agency; subsequently, NEPA established a broad national policy directing Federal agencies to maintain and preserve environmental quality. Reference publications related to environmental aspects of navigation projects are provided in EM.

NEPA requires all Federal agencies and officials to (a)direct their policies, plans, and programs to protect and enhance environmental quality;(b)view their actions in a manner that will encourage productive and enjoyable harmony between man and his environment;(c)promote efforts that will minimize or eliminate adverse effects to the environment and stimulate the health and well-being of man;(d)prompt the understanding of ecological systems and natural resources important to the nation; (e)use a systematic and interdisciplinary approach that integrates the ecological, social, cultural, and economic factors in planning and decision-making; (f)study, develop, and describe alternative actions that will avoid or minimize adverse impacts; and (g) evaluate the short-and long-term impacts of proposed actions.

Problems to be considered in the development or improvement of waterways for shallow-draft navigation include the potential adverse effects of the project on environmental quality. Some of the factors that could affect the environmental quality of a waterway are the following.

1.Excessive sedimentation

Bank erosion potential, adjacent land use practices, and general soil characteristics should be given consideration during site selection to prevent undesirable environmental effects from sedimentation and to minimize or eliminate the need for maintenance dredging. The need for reduction of bank slopes or other means of protection such as use of vegetation, gabions, or rock riprap to reduce the tendency for erosion from currents and waves should be considered. Old bendway cutoffs during construction are becoming more important as aquatic habitats. Such areas function effectively as sediment traps and may require special treatment to maintain their effectiveness as desirable aquatic habitats. Disposal areas located adjacent to the main stream or tributaries should be designed and operated such that the effluent meets appropriate Federal and State water quality standards for suspended sediment.

2.Resuspension of contaminants

Construction and maintenance dredging could cause the resuspension of contaminants. This is most likely to occur in waterways that have been used in the past as carriers of industrial, agricultural, or municipal wastes. Existing and past industrial and agricultural practices within the watershed should be examined and, if deemed necessary, appropriate sediment and water chemistry testing conducted to evaluate the potential impact of any resuspended contaminants on the aquatic environment.

3.Increased water temperature

Care should be taken to prevent the unnecessary removal of woody vegetation adjacent to the waterway. If such removal is necessity, it may be possible to remove such vegetation from only one side of the waterway so as to maximize the shading effect.

4.Water table effects and excavated material

Canalization and subsequent pooling of water behind a lock and dam may result in changes in the water table, thus changing the vegetation and the habitats available.

A major concern in many project areas will be the methods used to remove and treat excavated and dredged materials, depending on the nature of the materials and their potential for releasing contaminants.

5.Impacts on Aquatic, wetland, and terrestrial habitats

The route selected, construction activity, and management and operation of the project are all likely to have some adverse effects on biological habitats. The project and alternatives available should be evaluated to determine if any of the adverse effects could be eliminated or at least minimized. It might be possible to provide alternate habitats for certain species that are seriously affected.

6.Interruption of migratory routes

Evaluation of the use of the streams and adjacent terrestrial habitats as migration routes for aquatic and terrestrial animals is an important consideration during the planning process. Critical routes should be avoided when practical or provisions should be made for allowing alternate passage of the affected animals. Construction and maintenance activities could also be scheduled in such a manner as to avoid peak migration periods to reduce impact. Provision to pass migratory fish around or through the navigation structures are usually needed.

7.Modification of riparian habitats

Bottomland hardwood forests are regarded as an important, although rapidly disappearing, riparian habitat. Alternatives to the removal of existing natural riparian habitat should be developed so as to lessen such an adverse impact. Plans for revegetation should be developed where habitat modifications are necessary.

8.Disruptions of breeding or nursery areas

Certain areas such as Cypress or Tupelo swamps, marshes, and oxbow lakes along rivers and streams are more critical than others for breeding, nursery or nesting areas for aquatic, terrestrial, or arboreal animals. Particular care should be taken to identify such areas and arrive at suitable alternatives to the disruption of such habitat.

9.Increased turbidity

Turbidity is an indication of suspended and colloidal materials in the water. Continuing high turbidity levels in waterway over preproject conditions could adversely affect aquatic species. Measures such as construction of sediment traps, reseeding of construction areas, and construction of channel bypasses to prevent project contributions to increases in turbidity should be carefully considered in all phases of project design,

10.Impact on wetlands

Our nation 'wetlands have been diminishing rapidly during the past half century, such wetlands, in addition to serving as valuable habitat for diverse fish and wild lifte communities often are valuable for natural purification of polluted or contaminated waters. Wetlands also serve to eliminate severe changes in the water table, and often are highly regarded aesthetically. It may be possible, in proper consideration, to enhance wetland habitats along waterways and prevent unnecessary losses to existing wetland areas by using dredged material to create additional wetland areas.

11.Changes associated with the formation of bend way cutoffs

Many shallow-draft waterway projects result in the formation of bendway cutoffs by channelization for realignment of the navigation channel. Such areas in the past often served as a repository for excess dredged materials. This is no longer an acceptable practice and, furthermore, the potential value of such bendway cutoffs as aquatic habitat and recreation areas is frequently included in the planning and design. These areas are often subject to rapid sedimentation and filling by bedload materials, and some strict measures are often required to prevent the premature loss of these areas as aquatic habitats.

12.Ongoing corps of engineers studies

Extensive studies on the upper Mississippi and Illinois waterways are evaluating the effect of navigation on fisheries, aquatic plants, mussels, shallow water habitat, water quality, recreational resources, and historic properties.

New Words and Phrase

1.implementation *n.* 执行，实施
2.harmony *n.* 融洽；和睦
3.Stimulate *vi.* 促进；激发
4.interdisciplinary *n.* 跨学科的
5.interruption *n.* 阻断物
6.migratory *adj.* 迁移的
7.nursery *n.* 苗圃
8.turbidity *n.* 混浊度
9.colloidal *n.* 胶质的
10.gabion *n.* 石笼
11.disposal *n.* 清除；处理
12.effluent *n.* 污水
13.contaminant *n.* 污染物
14.deem *vi.* 认为；视为
15.canalization *n.* 渠道网
16.purification *n.* 提纯，净化
17.Premature *adj.* 未成熟的；过早的
18.mussel *n.* 蚌；贻贝

Notes

1.Problems to be considered in the development or improvement of waterways for shallow-draft navigation include the potential adverse effects of the project on environmental quality.

宽浅吃水航道的发展和提升工程所涉及的问题包括了实施工程后对环境质量的潜在不利影响。

draft 吃水　　waterway 水运，航道

The depth between LAT and the bottom of the harbor basin is determined by the draft of calling ships when fully loaded plus an overdepth to cover the trim of the ship, wave height and safety margin against sea bottom irregularities.

2.Canalization and subsequent pooling of water behind a lock and dam may result in changes in the water table, thus changing the vegetation and the habitats available.

船闸或者大坝修建所形成的运河和水塘将引起水位的改变，从而改变植被和可获得的栖息地。

result in　引起，后面一般接结果　　water table　水位

Businesses make large contributions to members of Congress, hoping to influence their votes on key issues.

3.These areas are often subject to rapid sedimentation and filling by bedload materials, and some strict measures are often required to prevent the premature loss of these areas as aquatic habitats.

这些区域一般会快速淤积或者被河床物质填满，因此需要开展一些严格的测量来避免这些水生态栖息地的损失。

be subject to　受……支配；受……影响

This method was used to build the piers for the Golden Gate Bridge in San Francisco, which is subject to strong tides and high winds, and is located in an earthquake zone.

Comprehensive Exercises

I.Answer the following questions on the text.

1.What are the factors to be considered in water project?

2.What is the influence of the maintenance dredge on the river environment?

3.What is the water table effects on the river environment?

4.How can we take action to improve the river environment?

5.How do the impact of the water project on the habitat of fish?

II.Fill the most appropriate words or phrases in the correct forms in the blanks from the list below.

implementation	enhance	habitat	dredge	adjacent to
conduct	so as to	available	subject to	indication

1.The test _____ at varying moisture contents in order to obtain a range of points along the compaction curve for the soil in question.

2.For any management system the benefits need to be at least equal to or greater than the cost of ______to ensure that the investment is worthwhile.

3.Its purpose is to increase the depth of water in a river______improve navigation.

4.In this technique a channel ______along the line of the tunnel; in other words, silt is pumped out of the water bed.

5.A civil engineer who does not know about new materials that______cannot compete successfully with one who does.

6.These waters have also formed ______for fish and wildlife in lands that were once nearly barren.

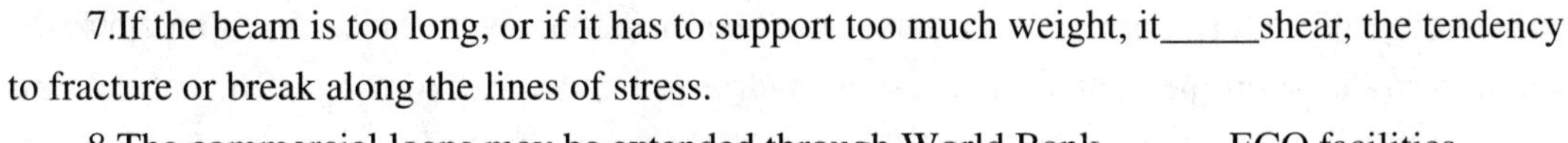

7.If the beam is too long, or if it has to support too much weight, it_____shear, the tendency to fracture or break along the lines of stress.

8.The commercial loans may be extended through World Bank ______ECO facilities.

9.When there is a heavy rainfall or rapid snowmelt, the river overflows into a wide flat area ______the channel.

10.To meet this requirement, tunnels must have walkways with handrails,_______of the direction of escape and distances and have sufficient and reliable emergency lighting.

III.Translate the following sentences into Chinese from the text.

1.Bank erosion potential, adjacent land use practices, and general soil characteristics should be given consideration during site selection to prevent undesirable environmental effects from sedimentation and to minimize or eliminate the need for maintenance dredging.

2.Existing and past industrial and agricultural practices within the watershed should be examined and, if deemed necessary, appropriate sediment and water chemistry testing conducted to evaluate the potential impact of any resuspended contaminants on the aquatic environment.

3.Many shallow-draft waterway projects result in the formation of bendway cutoffs by channelization for realignment of the navigation channel. Such areas in the past often served as a repository for excess dredged materials.

4.It may be possible, in proper consideration, to enhance wetland habitats along waterways and prevent unnecessary losses to existing wetland areas by using dredged material to create additional wetland areas.

Lesson 3 Hydroecology and Ecohydrology: Challenges and Future Prospects

The chapters clearly demonstrate the need for research at the intersection of hydrology and ecology to advance scientific understanding, develop novel methods, and manage the balance between growing human demands for water and needs of water-dependent ecosystems.Rather than a retrospective, we use this chapter to present a forward-looking view on the challenges and prospects for the ‘emerging discipline’ of hydroecology/ecohydrology and, in doing so, aim to focus attention on issues that require further evaluation and thought. We have suggested that a potential impediment to the development of ecohydrology/hydroecology is the lack of a clear subject definition to serve as a focal point to unite the research community. Most importantly, we assert herein that it is not simply the integration of hydrology and ecology that will determine the future prospects for ecohydrology/hydroecology but the way in which this integrative science is conducted. We advocate a truly interdisciplinary (as opposed to multi-disciplinary) approach in which ecologists and hydrologists benefit from genuine synergy by embracing advances at the cutting-edge of both sciences and a unified, focused methodological stance. Such an approach should provide more perceptive answers to hydroecological/ecohydrological problems and management questions. This context end by identifying future challenges and potential research themes that hydroecology/ecohydrology need to address as the discipline moves forward into the future.

1.The need for an interdisciplinary approach

The need for, and potential benefits of, research that bridges the gap between traditional discipline boundaries is generally well recognised by scientists, those funding pure, strategic and applied research(e.g., NERC, 2007; Environment Agency, 2004), and organisations that create and implement legislation relating to water (e.g., EU Water Framework Directive). A single discipline focus may neglect parallel concepts and key process interactions, result in research undertaken at inappropriate (too fine or too coarse) or disconnected (mismatched) scales, and/or fail to make use of the most powerful, cutting-edge investigative tools. This wider realisation of the need for 'hybrid science' and interdisciplinary approaches has certainly stimulated much recent interest in hydroecology. Although research focusing upon the interaction between hydrology and ecology has existed for some time, true integration (especially between linked and coupled hydrological and ecological processes) remains incomplete. Arguably, this may be related to the way in which such research has been practised. Thus, the logic for adopting a hydroecological approach is persuasive but the way that hydrology and ecology are being integrated requires further careful evaluation. At present, ecologists and biologists appear to be looking at research questions from one perspective and hydrologists (mainly geographers and engineers) from another. Scientists may be seeking to address the same issue or solve the same problem without converging on the most perceptive or robust hydroecological answer(s) due to the absence of a clear theoretical understanding of the key elements in the 'other' discipline. If a true paradigm shift is to occur and hydroecology is to flourish, ecologists and hydrologists need to bridge the gap between traditional subject boundaries to build real interdisciplinary teams and so reap benefit from true synergy by embracing advances at the cutting-edge of both sciences.

Hence, we see a clear need for a genuine interdisciplinary approach; and we consider this a vital first step for generating new insights into water-ecosystems interactions and, in turn, identifying the future research agendas. To provide fully integrated solutions to issues affecting waterdependent systems, it may be necessary for hydrologists and ecologists to expand their interdisciplinary teams even further to include, for example, atmospheric scientists (e.g.,climate change), geologists (e.g., groundwater-surface water interactions), biogeochemists (e.g., nutrient availability), soil scientists (e.g.,material properties influencing infiltration and soil biota activity), and social scientists(e.g., socio-economic impacts) to understand the full cascade of processes and their mechanistic interactions.Making a plea for interdisciplinarity is reasonably straightforward (may be even somewhat trite) but we recognise that building a framework that fosters interdisciplinary collaboration between hydrologists and ecologists is no simple task. There is a need to go beyond mere 'discipline- hopping' (in terms of subject areas) and to meld reductionist Newtonian (often adopted by physical scientists-based upon simplification, universal laws and predictive understanding) and holistic Darwinian (often adopted by ecologists-based upon principles of complex interdependency) philosophies. This split between Newtonian hydrologists and Darwinian ecologists is probably fuzzier in actuality.

Harte identifies 'elements of synthesis' for Newtonian and Darwinian traditions in the context of earth systems science as: (i) simple, falsifi able models,(ii) search for patterns and laws, and (iii) embracing the science of place. Most notably,the major scientific benefits could be gained from combining reductionist and holistic systems approaches. Hierarchical scaling theory may offer a means of spanning the gap between these two approaches. Within this framework, (reductionist) explanations for phenomena/models can be considered in terms of their significance at different levels of organization (i.e., in a more holistic sense). Newman propose that by examining adjacent levels in the hierarchy, new discoveries at the higher or lower scale may be made. This nested, coupled approach is not 'new' as such a hierarchical view has already been embraced successfully by some hydroecological/ecohydrological studies.Place-based research may provide another framework for developing hydroecological collaboration with focus upon: (sub-)system based studies, integrated basins studies or a network of (long term) monitoring sites. Newman identified three crosscutting problems that represent barriers to overcoming key challenges in plant- water interaction research but they are applicable more generally to place-based hydroecological research, and also how results can be used to predict response/change and manage the environment.

2.Future research themes

The needs for both a unifying definition of hydroecology and a framework to promote true interdisciplinary approach are generic, structural, challenges to the future of hydroecology. The examples highlighted from fluvial and riparian ecosystems are indicative, demonstrating the highly interrelated nature of hydroecological research, although we recognise that similar challenges apply to other environments and ecosystems such as drylands and wetlands.

2.1 Ecosystem sensitivity to hydrological change

An increasing number of ecosystems have been identified as being vulnerable to hydrological change. However, while this vulnerability is recognised, knowledge is lacking regarding the most ecologically sensitive periods to hydrological extremes (e.g., floods, low flows/droughts and soil moisture deficit), and associated water stress and habitat disturbance. In addition to focusing upon extremes, there is an evolving awareness of the importance of considering the spectrum of hydrological conditions experienced by habitats, and their linkages to ecosystem structure and functioning. Long-term data sets have a critical role to play in unravelling these hydroecological associations and setting short-term change/variability in a wider context.

2.2 Disturbance: water and ecological stress

Disturbance is widely accepted as one of the primary driving forces behind ecosystem change. Most natural disturbances (e.g., floods, drought and fire) occur as a result of a climatological aberration, which propagates down the atmosphere- hydrosphere-land surface-biosphere cascade. Superimposed upon natural (climate) disturbances, anthropogenic activities have the potential to enhance or moderate impacts significantly. However, the role of predictability and stochasticity in natural disturbance processes has not received the attention warranted (Plachter and Reich, 1998). For example, floodplains are highly dynamic environments that are in a state of constant flux with repeated erosion and sedimentation processes, and inundation and desiccation events. Thus, the riparian habitat is strongly influenced by channel kinetics, determined predominantly by the frequency and magnitude of flood events. This (often seasonal) flood disturbance is thought to maximise biological processes, and both in-stream and riparian biodiversity. In systems where floods are less predictable and 'reset' the physical and biotic environment, they can have catastrophic outcomes for populations of organisms, even though many have adaptations for dealing with inundation. Urgent research foci include: (i) quantifying variability;(ii) understanding disturbance cause (hydrology) and effect (ecology) and feedbacks; and(iii) predicting disturbance regimes under changing climate and/or with growing human impact.

2.3 Aquatic–terrestrial linkages

Aquatic ecosystems are strongly coupled to their terrestrial counterparts as they receive water via overland flow, shallow and deep subsurface flow paths and in some cases direct precipitation. A number of long-standing ecological concepts and theories have been advanced to explain how riverine ecosystems interact with the landscape and affect ecosystem processes. The balance between the supply of autochthonous and allochthonous organic materials from headwaters and floodplains and its importance for riverine energetics, nutrient supply and foodwebs is a fundamental tenet of the three key lotic ecosystem models: the river continuum concept (RCC), the flood pulse concept (FPC) and the riverine productivity model (RPM) (Thorp and Delong, 1994). The FPC postulated that in unaltered large river systems, the bulk of riverine animal biomass relates directly or indirectly to production in the floodplains and not downstream transport of material from elsewhere (cf. RCC). Set within these large-scale theories, there are finer grain linkages characterised by smaller scale 'foodweb subsidies' that move across terrestrial, aquatic and even marine habitats.

A growing number of studies have highlighted the significance of terrestrial arthropods as essential food resources in aquatic stream ecosystems and recent work has shown how aquatic biodiversity can support certain riparian organisms. The range of spatio-temporal variability in these foodweb subsidies is not fully quantified at present and, hence, represents significant opportunity for new hydroecological research.

2.4 Modern and palaeo-analogue studies

The potential value of long-term hydroecological datasets has received extensive comment . Aside from data generated by routine monitoring, such as annual surveys collected by organisations responsible for managing habitats (e.g., the statutory national environmental monitoring agencies), longterm paired hydrological and ecological datasets that span more than a decade are uncommon. Long-term data are increasingly regarded as important for understanding larger temporal fluctuations of aquatic communities that result from both anthropogenic activities and natural variability.

In this regard, palaeoecological studies are valuable under-utilised resources. There is a wealth of palaeolimnological research that shows how data derived from subfossil organisms (e.g., diatoms and Cladocera) can be used as baseline information to understand longer-term hydroecological variability. Although rather too infrequently applied to other aquatic systems, this palaeoecological approach has demonstrated potential to reveal hydrological change in river basins, and ecological patterns and processes that are associated with such change.

2.5 Applied hydroecology

Balancing anthropogenic demands for water against the water needs of aquatic ecosystems is a pressing global issue. To achieve this goal requires collaboration between academics (e.g., hydrologists, geomorphologists, engineers and ecologists), practitioners (e.g., water and habitat managers) and stakeholders (e.g., landowners, anglers and recreational users) to balance the multiple and often conflicting pressures associated with the management of the system (e.g., flood alleviation and reservoir management) against the protection, and even enhancement, of ecosystem properties (e.g., conservation and restoration of habitats). River restoration is a good illustration of these issues and arguably represents one of the greatest challenges for hydroecologists.

In many instances, the goals and standards for historic restoration projects have arguably been poorly defined. Consequently, there has been a mismatch (even conflict) between what stake holders would like, what practitioners can deliver with available funds, and scientists'recommendations for environmental and ecological benefit. This has been paralleled by an increase in small scale restoration projects (e.g., at the scale of individual riffles or flow deflectors to enhance local flow velocity) that may be hydraulically and morphologically enhanced but result in minor or insignificant effects on instream communities. This may reflect the disconnected nature of the individual projects in relation to the wider river basin context and also the need to identify the appropriate scale at which to undertake restoration. It would seem that clearly defined pre-project aims and standards are required that integrate scientists, practitioners'and users' perspectives; post-project appraisal is also required using multiple indicators. Scientists (hydrologists and ecologists together), environmental managers and stakeholders faced with hydroecological problems require to converge upon a common vision and a unified approach.

New Words and Phrase

1.retrospective *adj.* 回顾的；涉及以往的
2.impediment *n.* 妨碍，阻碍
3.assert *v.* 明确肯定，断言
4.integrative *adj.* 综合的，整体的
5.synergy *n.* 协同作用
6.stance *n.* 观点，态度
7.legislation *n.* 法规，法律
8.persuasive *adj.* 有说服力的
9.robust *adj.* 强健的，强壮的
10. paradigm *n.* 典范，范例
11. agenda *n.* （会议的）议程表，议事日程
12. cascade *v.* 倾泻，流注
13. hierarchical *adj.* 按等级划分的
14. vulnerable *adj.* 脆弱的，易受……伤害的
15. autochthonous *adj.* 土生土长的，原地形成的
16. palaeoecological *adj.* 原生态的
17. stakeholder *n.* （某组织、工程、体系等的）参与人
18. appraisal *n.* 评价；估价

Notes

1.We advocate a truly interdisciplinary (as opposed to multi-disciplinary) approach in which ecologists and hydrologists benefit from genuine synergy by embracing advances at the cutting-edge of both sciences and a unified, focused methodological stance.

我们提倡真正的跨学科的（而不是多学科的）方法，使得生态学家和水文学家都能从科学前沿和统一标准的方法论观点出发，通过协同合作获得所需要的研究利益。

benefit from 得益于；得利于；因……而得到好处

cutting-edge 前沿，尖端的

Taxpayers received little benefit from these cuts because many of the jobs that remained are among the highest paid in the federal government.

2.In addition to focusing upon extremes, there is an evolving awareness of the importance of considering the spectrum of hydrological conditions experienced by habitats, and their linkages to ecosystem structure and functioning.

除了极端条件的研究外，研究者还需要逐渐意识到栖息地的水文条件范围的重要性，以及它们和生态结构及运行的联系。

awareness of 意识到

The solution is to develop an awareness of how thoughts and behaviors at the individual level affect thoughts and behaviors at the global level.

3.This may reflect the disconnected nature of the individual projects in relation to the wider river basin context and also the need to identify the appropriate scale at which to undertake restoration.

这反映了宽广流域环境内工程的不连续性本质以及需要确定合适的尺度来进行完成对其的修复工程。

in relation to　关于，涉及

Piers, where visible, should not appear either massive or than in relation to deck thickness.

Comprehensive Exercises

I.Answer the following questions on the text.

1.What are the challenges for the hydroecology and ecohydrology?

2.How do we find the approach to bridge the gap between traditional discipline boudnaries?

3.What are the research themes in the future prospect?

4.What are the primary driving forces behind ecosystem change?

5.How do we balance anthropogenic demands for water against the water needs of aquatic ecosystems?

II.Fill the most appropriate words or phrases in the correct forms in the blanks from the list below.

aim to	forward	make use of	search for	demonstrate
vulnerable	propagate	aside from	appraisal	converge

1.A large cylinder with a cutting edge that can be moved_____by jacks.

2.Pumping stations are one of the more ______features of a flood protection project.

3.This paper is ______investigate the influence of the dam construction on the ecological environment in the backwater area.

4.If people ______an idea or piece of information, they spread it and try to make people believe it or support it.

5. _____the water availability in the topsoil, the evaporation from a cropped soil is mainly determined by the fraction of the solar radiation reaching the soil surface.

6.The temporary nature of the project means that investments in product quality improvement, especially defect prevention and ______.

7.Therefore it becomes increasingly important to consider carefully what is the appropriate level of water quality to______alternative ways of improving water quality.

8.An iterative method_____on one of the solutions, without any indication of the existence of the other ambiguous solution.

9.This article provided some descriptions of the API, along with some sample code to illustrate how to ______them.

10.On—the—on training can be acquired that _____his or her ability to translate theory into practice to the supervisors.

III.Translate the following sentences into Chinese from the text.

1.Rather than a retrospective, we use this chapter to present a forward-looking view on the challenges and prospects for the 'emerging discipline' of hydroecology/ecohydrology and, in doing so, aim to focus attention on issues that require further evaluation and thought.

2.If a true paradigm shift is to occur and hydroecology is to flourish, ecologists and hydrologists need to bridge the gap between traditional subject boundaries to build real interdisciplinary teams and so reap benefit from true synergy by embracing advances at the cutting-edge of both sciences.

3.Disturbance is widely accepted as one of the primary driving forces behind ecosystem change. Most natural disturbances (e.g., floods, drought and fire) occur as a result of a climatological aberration, which propagates down the atmosphere- hydrosphere-land surface-biosphere cascade.

专业论文英文写作方法分析

科技英语论文的基本组成包括：题名、摘要、正文、致谢、参考文献。

1. 题名

科技论文的题名是表达论文的特定内容，反映研究范围和深度的最恰当、最简明的逻辑组合，应以最精炼的文字，充分表述论文的内容。

1.1 题名的要求

（1）准确。即题名能够准确地反映论文的主要内容。

（2）简练。即题名应当言简意赅，以最少的文字概括尽可能多的内容。

（3）醒目。即题名要清晰地反映论文的特色，明确表明研究工作的独到之处，甚至要有一定的生动性和新颖性。

1.2 题名撰写的基本要求

（1）题名通常由名词性短语构成，其中的动词多以分词或动名词形式出现。题名不应由陈述句构成，偶尔使用探讨性语气的疑问句作为题名，例如：When is a bird not a bird?

（2）为突出论文的核心内容，应尽可能将表达核心内容的主题词放在题名开头，以便引起读者的注意。

（3）题名中常可以删去不必要的冠词（a，an，the）及多余的说明性“冗词”。例如：Study/Analysis/Development/Evaluation/Investigation/Review of…,The effect of…,Report of/on…,Research on…,A Discussion/Observations on…,Preliminary study/Discussion of…,The

Preparation/synthesis/nature of…,Treatment/Use of…。

（4）注意题名中介词 with，of，for，in 的用法。

（5）题名中慎用缩略语，只有那些全称较长，已得到科技界公认的缩略语，才可使用。缩略语全部用大写字母，如：LASER，AIDS，CAD，GIS 等。

（6）特殊字符，如数学符号和希腊字母在题名中尽量不用。

2. 摘要

摘要是以提供文献内容梗概为目的，不加评论和补充解释，简明、确切地记叙文献重要内容的短文。

2.1　摘要的要素

（1）目的：研究工作的前提、目的和任务，所涉及的主题范围。

（2）方法：所用的理论、条件、材料、手段、装备、程序等。

（3）结果：观察、实验的结果，数据，得到的效果，性能等。

（4）讨论：结果的分析、比较、评价、应用，提出的问题，今后的课题，假设，启发，建议，预测等。

2.2　摘要撰写的基本要求

（1）摘要中可适当强调研究中的创新、重要之处，但不要使用自我评价性语言，例如：……具有重要意义、可观的推广应用前景、为……提供依据等。摘要应尽量包括论文中的主要论点和重要细节（重要的论证或数据）。

（2）使用简短的句子，用词应为潜在的读者所熟悉。表达要准确、简洁、清楚。注意表达的逻辑性，尽量使用指示性词语表达论文的不同部分（层次），如使用“We found that…”表示结果，使用“We suggest that…”表示讨论的结果。

（3）时态。叙述研究过程（包括实验、观测、数据处理等）用一般过去时；阐述研究结果、结论、叙述客观事实等用一般现在时，表示该文“报告”“描述”及“讨论”等意思时也用一般现在时。一般不用完成时、进行时和其他复合时态。

（4）语态。谓语动词尽量使用主动语态，少用被动语态，但在叙述研究过程时可用被动语态使重点突出。例如：this paper presents formulas 和 formulas are presented，前者简洁、有力，后者突出重点。

（5）代词。以客观事实作主语，尽量不用人称代词，少用第一、第二人称代词。必要时用动名词作主语。

（6）句型。尽量用简单句来表达，即使有必要使用复合句型，也要尽量使用简单复合句，有些从句可改用分词短语。

（7）删繁就简，能用一个词则不要用词组，能用简单词则不要用复杂词。例如：用 because 不用 due to，用 to 不用 in order to，用 use 不用 utilize。

（8）可用动词的情况尽量不用其名词形式，避免使用 be，have，do 等弱动词。例如：用 thickness of plastic sheets was measured，而不用 measurement of thickness of plastic sheets was made。

（9）避免使用长系列形容词或名词来堆列修饰名词，可用预置短语分开，或用连字符断开名词词组，作为组合形容词。例如：water saving irrigation agriculture 改用 agriculture of water-saving irrigation。

（10）文辞要朴实无华，不用文学性描述。尽量简化一些措辞和重复的单元。例如：用 at 250-300℃，不用 at a temperature of 250℃ to 300℃；用 at 200Pa，不用 at a high pressure of 200Pa；用 at 1500℃，不用 at a high temperature of 1500℃。

2.3　摘要的行文方法

1）研究目的的表达方法

（1）主动句型

This paper describes/presents/discusses/analyses/reports on/investigates/examines/deals with/researches into/gives/points out/reviews/... 本文描述 / 提出 / 讨论 / 分析 / 报告 / 调查 / 检验 / 论述 / 探讨 / 给出 / 指出 / 总结 /……

This paper is concerned with/aimed at/limited to/related to… 本文研究 / 旨在 / 限于 / 关于……

The purpose (aim/objective/…) of this paper is to discuss/study/research/… 本文目的是……讨论 / 研究 / 探讨 /……

也可使用名词句型：This paper makes a study of... /This paper makes investigations on…

（2）被动句型

Information regarding... is described/presented/discussed/analyzed/reported/ investigated/ examined/dealt with/given/pointed out.

In this paper a new method (approach) of... is introduced (recommended) . 本文介绍的一种……的新方法。

By using… , 使用……；…was studied，对……进行了研究。

2）研究方法的表达方法

An example of… is analyzed/described/discussed/examined/studied in detail. 详细地分析 / 描述 / 讨论 / 检验 / 研究 / 了……的例子。

Data are displayed in graphs and tables. 数据显示在图表中。

Findings（Results）are presented/reported/analyzed/examined/discussed. 提出 / 报告 / 分析 / 检验 / 讨论了新发现（结果）。

A series of experiments were made/carried out on… 对……进行了一系列实验。

Special mention is given here to… 这里专门提到……

Examples of… benefit from that…　……的例子表明……

Statistics confirm…　统计数字肯定了……

3）研究结果的表达方法

Facts show that…　事实证明……

Experiment finds that…　实验表明……

Study proves that…　研究证明…

Results show/indicate/reveal/suggest/illustrate/demonstrate/…　结果证明/表明/揭示/提出/说明/表明/……

Comparison concludes that…　这一比较推断出……

Statistical analysis demonstrates that…　统计分析提出……

根据摘要内容的不同，可分为以下几种表达方法：

（1）说明理论、分析方法或设计方法

The result of this study can be generalized/finalized for…　这一研究成果可推广到/概括……

Acceptable results of design were obtained by the method for…　通过……方法，获得了在设计上可以接受的结果。

The results of this study are summarized/summed up as follows/in the following. 研究结果归纳如下。

（2）说明使用数学模型和公式结果

Calculations made with this formulation show that…　用这一公式计算表明……

An exact expression is obtained and the results are analyzed. 获得了一项正确公式，并且分析了其结果。

（3）说明实验或其他方面结果

The conclusions were drawn from the test results. 从试验结果得出这些结论。

Results for… are found to be close to the experimental data. 已经证明……的结果与实验数据相接近。

The results are illustrated by a specific example. 其结果可由一具体实例加以说明。

4）摘要尾句的表达方法

These results/data/findings/experiments also indicate that…　这些结果/数据/发现/实验/已指出……

The findings imply that…　这些发现暗示……

Based on these conclusions… is discussed. 根据这些结论，讨论了……

The findings suggest that further research into … is called for（would be worthwhile）这些发现提示对……应进一步研究（对……进一步研究是值得的）

3. 正文

正文是科技论文的主体部分，包括引言、材料与方法、结果、讨论、结论。

3.1 引言

引言具有总揽论文全局的作用，是论文中最难写的部分。引言的基本内容包括研究背景、存在问题和研究目的三个方面。

3.1.1 引言撰写的基本要求

（1）研究背景的阐述要繁简适度。

（2）在背景介绍和问题的提出中，应引用“最相关”的文献以指引读者。优先引用的文献包括相关研究中经典的、重要的和最具说服力的文献，力戒刻意回避引用最重要的文献，甚至是对作者研究具有“启示性”的文献，或者不恰当地大量引用作者本人的文献。

（3）提出的研究问题包括以下几种形式：

①以前的学者尚未研究或处理得不够完善的课题。

②过去的研究衍生出有待探讨的新问题。

③以前的学者曾提出两个以上互不相容的理论或观点，必须做进一步研究，才能解决这些冲突。

④过去的研究很自然可以扩展到新的题目或领域，或以前提出的方法或技术可以改善或扩展到新的应用范围。

（4）采取适当的方式强调作者在本次研究中最重要的发现或贡献，让读者顺着逻辑的演进阅读论文。切忌故意制造悬念，以期望在论文的最后达到高潮，甚至将重要的发现在摘要中也忽略。实际上这种做法往往适得其反，因为读者不一定有耐心阅读冗长的文字，直至坚持到最后的重要部分。

（5）解释或定义专门术语或缩写词，以帮助读者阅读与理解。

（6）适当地使用 I、We、Our，以明确地指示作者本人的工作，例如：最好使用 We conducted this study to determine whether…，而不使用 This study was conducted to determine whether…。

（7）叙述前人工作的欠缺以强调自己研究的创新时，应慎重且留有余地，切忌评价式用语，例如“首次发现、首次提出、达到国际先进水平”等。可采用类似如下的表达：To the author's knowledge…；There is little information available in literature about…；Until recently, there is some lack of knowledge about… 等。

3.1.2 引言的时态运用

通用的规则是：介绍已有的认识时，使用现在时；叙述本人或他人近期的工作或认识时，使用过去时。

1）研究背景

（1）介绍一般性资料、现象或普遍事实时，句子的主要动词多使用一般现在时。例如：Acid rain is a serious problem in many areas of Europe.

（2）引述其他学者过去的研究行为时多采用一般过去时，且常以 that 从句叙述被引作者的研究结果，而从句中动词的时态因所表达资料的性质而定。例如：Chen showed(found/reported/noted/suggested/observed/pointed out)that the water boils at 100℃.（从句中的资料为普遍事实，用现在时）

Chen found (reported/suggested/observed) that reducing the amount of oxygen caused the deposition rate to drop sharply.（从句中的资料尚未被视作普遍事实，用过去时）

如果 that 从句中的资料不是很确定的研究结果（如建议、假设等），则主句中的动词应使用 suggested, hypothesized 之类的臆测动词，从句中则使用 may ＋现在时。例如：

Ross suggested（hypothesized/proposed/argued）that reducing the duration of school vacations may help children to retain more of what they learn in class.

（3）描述特定研究领域中最近的某种趋势，或者强调某些最近发生的事件对现在的影响时，使用现在完成时。例如：In recent years, a variety of standards have been proposed in the literature.

2）存在问题

（1）如果叙述的是普遍事实，用现在时。例如：Little is known about X 或 Little literature is available on X.

（2）如果描述过去已开始并持续到现在的趋势或事件，用现在完成时。例如：Few studies have been done on X 或 Little attention has been devoted to X.

3）研究目的

（1）用 paper, report, thesis, dissertation 等表示论文提供资料的行为，重点在于介绍新的技术或方法、分析某个问题或提出某个论证。由于论文提供资料的行为是不受时间影响的事实，所以通常使用一般现在时。例如：

The purpose (aim/objective) of this paper is to analyze the effect of X on Y.

This paper presents (reports/describes/discusses) the results of experiments in which X was mixed with Y.

（2）采用 study, research, investigation, experiment 等介绍研究活动，重点在于提出某个调查或实验结果。由于句中所涉及的是已经结束的事情，因而多使用一般过去时。

In this research (study), we investigated the effects of X.

（3）由于所涉及的资料“将要”在论文中被提出来，因此有些作者也偏向使用将来时（句子中有 purpose, aim, objective 等名词时除外）。例如：

This paper will propose (present, evaluate, discuss) a new method for analyzing X.

The aim of this paper is to give.../The main purpose of the experiment reported here was to... / The study was designed to evaluate... / The primary focus of this paper is on... / The aim of this investigation was to test...

（4）有时作者在引言中还可较谦虚地，或者试探性地指出自己研究的价值，其中常用的助动词有 may，should，could 等。例如：

These findings may be useful to researchers attempting to increase employee productivity.

3.2 材料与方法

3.2.1 材料与方法撰写的基本要求

（1）对材料的描述应清楚、准确。

通常先对材料做概述，然后再详细描述材料的结构、主要成分或重要特性、设备的功能等。

（2）对方法的描述要详略得当、重点突出。

方法即描述研究是如何开展的？通常按研究步骤的时间顺序描述方法，如果没有时间顺序，则按重要性程度描述研究步骤。

（3）力求语法正确、表达简洁且合乎逻辑。

3.2.2 材料与方法的时态与语态运用

（1）若描述的内容为不受时间影响的事实，采用一般现在时。

（2）若描述的内容为特定的、过去的行为或事件，采用过去时。

（3）材料与方法章节的重点是描述实验进行的步骤及采用的材料，由于所涉及的行为与材料是讨论重点，读因而一般采用被动语态。但是，如果涉及表达作者的观点或看法，则采用主动语态或不定式结构。

3.3 结果

结果通常包括结果的介绍、结果的描述、对结果的评论。有些期刊常将结果与讨论合并。

3.3.1 结果撰写的基本要求

（1）对实验或观察结果的表达要高度概括和提炼。

（2）数据表达可采用文字与图表相结合的形式。

如果数据较多，可采用图表形式来完整、详细地表述，文字部分则用来指出图表中资料的重要特性或趋势，切忌在文字中简单地重复图表中的数据，而忽略叙述其趋势、意义以及相关推论。

（3）文字表达应准确、简洁、清楚。

避免使用冗长的词汇或句子来介绍或解释图表，避免把图表的序号作为段落的主题句，应在句子中指出图表所揭示的结论，并把图表的序号放入括号中。例如：

（劣）Figure 1 shows the relationship between A and B.

（优）A was significantly higher than B at all time points checked(Fig.1).

表达“比较”时，避免使用 compared with，应直接指出比较的结果。例如：

（劣）X was significantly increased compared with Y(Fig.1).

（优）X was significantly higher than Y(Fig.1).

3.3.2　结果的时态运用

（1）结果的介绍

即指出结果在哪些图表中列出，常用一般现在时。例如：

As Figure 2 shows, the temperature increased rapidly.

（2）重要结果的描述

叙述或总结研究结果的内容是关于过去的事实，常采用过去时。有时也采用现在时描述结果，差别是：使用现在时表示“该结果是在研究过程中所揭示的是普遍事实”；使用过去时表示“这是我们在本次研究中在某些特定情况下所发现的事实”。

（3）对结果的评论或说明

通常使用一般现在时。

3.4　讨论

讨论的基本内容包括：回顾研究目的、概述重要结果、说明结果或阐述相关推论、研究方法或结果的局限性及相关建议、结果的理论意义或实际应用。通常根据需要可将讨论与结论合并为“Discussion and Conclusion”或“Conclusion”，有些论文则将结果与讨论合并为“Result and Discussion”，其后为“Conclusion”。

3.4.1　讨论撰写的基本要求

（1）对结果的解释要重点突出，简洁、清楚。

为了有效地回答所研究的问题，可适当地、简要地回顾研究目的并概括主要结果。

（2）推论要符合逻辑，避免实验数据不足以支持的观点和结论。

（3）观点或结论的表述要清楚、明确。

尽可能清楚地指出作者的观点或结论，并解释其支持还是反对已有的认识。此外，要大胆地讨论工作的理论意义和可能的实际应用，清楚地告诉读者该项研究的新颖性和重要之处。

（4）对结果科学意义和实际应用效果的表达要实事求是，适当留有余地，避免使用“for the first time”等类似的优先权声明。

3.4.2　讨论的时态运用

1）回顾研究目的

通常使用过去时。

2）概述重要结果

如果作者认为所概述结果的有效性只是针对本次特定的研究，采用过去时；如果具有普遍的意义，采用现在时。

3）说明结果或阐述相关推论

（1）说明结果时多采用主从复合句的形式，主句动词多为表示可能性的现在时动词，从句中的动词为现在时表示说明具有普遍性意义，从句中的动词若为过去时，则说明的结果范围只适于本次的特定研究。例如：It is possible (may be/is likely) that adding water causes the reaction rate to increase.

（2）阐述由结果得出的推论时，通常使用现在时。

4）研究方法或结果的局限性及相关建议

（1）指出研究局限时，采用的时态视句子的内容确定。如果是关于作者已完成的研究事实，采用过去时。如果作者要指出自己的方法、模型或分析的局限，则采用现在时。例如：

Only two sets of conditions were tested.

Our findings may be only valid for females.

（2）建议新的题目或进一步研究时，常使用现在时，有时在动词前加语态动词would, could或表示更强烈建议的should。例如：

Experiments similar to those reported here should be conducted using different age groups.

偶尔，作者会提及自己正在进行或拟将进行的相关研究，这时可用现在进行时或将来时，而且最好用第一人称作主语，以便读者了解这是作者自己的行为，而不是在提建议。

5）结果的理论意义或实际应用

表述结果的理论意义或实际应用时，多采用现在时，并辅以may, might或should(should表示作者对自己研究的应用价值非常肯定）。例如：

Our findings may be useful to educators and others involved in curriculum development.

3.5　结论

通常情况下，结论包含在“结果与讨论”或“讨论”中，但也可将“结论”单列。结论的基本内容包括：

（1）作者本人研究的主要认识或论点，其中包括最重要的结果、结果的重要内涵、对结果的说明或认识等。

（2）总结性地阐述本研究结果可能的应用前景、研究的局限性及需要进一步深入的研究方向。

结论的时态运用与讨论的时态运用基本相同。

4. 致谢

1）致谢的基本内容：

（1）感谢任何个人或机构在技术上的帮助，其中包括提供仪器、设备或相关实验材料，协助实验工作，提供有益的启发、建议、指导、审阅，承担某些辅助性工作等。

（2）感谢外部的基金帮助，如资助、协议或奖学金，有时还需附注资助项目号、合同书编号等。

2）致谢撰写的基本要求

（1）致谢的内容应尽量具体、恰如其分。

致谢的对象应是对论文工作有直接和实质性帮助、贡献的人或机构，因此，致谢中应尽量指出相应对象的具体帮助与贡献。

（2）用词要恰当。

致谢的开始就用 We thank，不要使用 We wish to thank, We would like to thank, The authors thank 等。单词 wish 最好从致谢中消失，当表达愿望时，wish 是很好的词，但如果说 I wish to hank John Jones，则是在浪费单词，且可能蕴含着“我希望感谢 John Jones 的帮助，但这种帮助并不那么大”。实际上用 I thank John Jones，显得更为简明和真诚。

5. 参考文献

选择参考文献的基本原则：

（1）所选用文献的主题必须与论文密切相关，可适量引用高水平的综述性论文，以概括一系列的相关文献。

（2）必须是亲自阅读过，若为间接引用（即转引某篇论文的引文），则需要提及是从哪篇文献中转引的。

（3）尽可能引用已公开出版，而且最好是便于查找的文献，即同等条件下应优先引用著名期刊上发表的论文。

（4）尽量避免引用非公开出版物，私人通信的方式应遵照拟投稿期刊的习惯或相关规定处理。

（5）优先引用最新发表的同等重要的论文。

（6）一般不引用专利和普通书籍（如大学本科生教材等）。

（7）避免过多地，特别是不必要地引用作者本人的文献。

（8）确保文献各著录项（作者姓名，论文题目，期刊或专著名，期刊的年、卷、期或专著的出版年、出版地、出版社，起止页码等）正确无误。